Gamma Rays - Current Insights

Edited by Hosam M. Saleh and Amal I. Hassan

Published in London, United Kingdom

Gamma Rays - Current Insights
http://dx.doi.org/10.5772/intechopen.111057
Edited by Hosam M. Saleh and Amal I. Hassan

Contributors
Hosam M. Saleh, Amal I. Hassan, Godswill Ntsomboh Ntsefong, Fokam Paul Ernest, Likeng-Li-Ngue Benoit Constant, Tabi Mbi Kinsley, Zambou Alain Hervé, Mafouasson Hortense Noelle, Bell Joseph Martin, Imaizumi Masayuki, Kalpana Tewari, Mahipal Singh Kesawat, Vinod Kumar, Chirag Maheshwari, Veda Krishnan, Sneh Narwal, Sweta Kumari, Anil Dahuja, Santosh Kumar, Swati Manohar, Hussein Farid Mohamed, Samira Elsayed Mustafa El-Naggar, Mahmoud Abd-elmohsen Sweilem, Ahmed Adly Mahmoud Ibrahim, Ola Elsayed Abd Alrahman El-khawaga

First published in London, United Kingdom, 2024 by IntechOpen
IntechOpen is the global imprint of INTECHOPEN LIMITED, registered in England and Wales, registration number: 11086078, 5 Princes Gate Court, London, SW7 2QJ, United Kingdom

British Library Cataloguing-in-Publication Data
A catalogue record for this book is available from the British Library

Additional hard and PDF copies can be obtained from orders@intechopen.com

Gamma Rays - Current Insights
Edited by Hosam M. Saleh and Amal I. Hassan
p. cm.
Print ISBN 978-0-85466-383-5
Online ISBN 978-0-85466-382-8
eBook (PDF) ISBN 978-0-85466-384-2

Meet the editors

Hosam Saleh is a Professor of Radioactive Waste Management at the Radioisotope Department, Atomic Energy Authority, Egypt. He was awarded an MSc and Ph.D. in Physical Chemistry from Cairo University, Egypt. Dr. Saleh has more than 25 years of experience in hazardous waste management with an emphasis on the treatment and development of new matrixes for the immobilization of these wastes. He is also interested in studying innovative economic and environmentally friendly techniques for the management of hazardous and radioactive wastes. He has authored many peer-reviewed scientific papers and chapters and served as editor of several books. He was listed among the top 2% of scientists in the world by Stanford University, USA, in 2020, 2021, 2022, and 2023.

Amal Hassan is a Professor of Animal Physiology in the Department of Radioisotopes, Nuclear Research Center, Atomic Energy Authority, Egypt. She has authored many peer-reviewed publications on chronic diseases and is an editor, reviewer, and referee for several international scientific journals. She obtained a Certificate of Excellence in international scientific research arbitration from Publons. She was selected for inclusion in the global encyclopedia *Who's Who* in 2014, 2015, and 2016.

Contents

Preface

Gamma rays have become an indispensable tool in diverse fields of science and technology in the modern era. Their unique properties allow gamma rays to be utilized innovatively to advance scientific knowledge and drive technological progress. Therefore, the scientific community must continue to explore novel applications of gamma rays and share findings on their utilization. This book extensively examines the various domains in which gamma rays are significant, highlighting their promise in areas such as geological surveying and agricultural progress. The chapters in this volume seek to offer readers a thorough comprehension of the diverse applications of gamma rays, encompassing both theoretical and practical elements.

The opening chapter establishes the context for our exploration, encouraging readers to explore the fascinating realm of gamma rays and its complex links to scientific exploration and technological advancement. It functions as an entry point to the following chapters, briefly previewing the extensive range of opportunities that gamma rays offer.

Chapter 2 explores the field of radioactive prospecting, emphasizing the effectiveness of gamma-ray spectroscopy in identifying fault traces. This application demonstrates the crucial significance of gamma rays in geological investigation, enhancing our comprehension of the Earth's composition and structure.

Chapter 3 delves into the use, progress, and difficulties of gamma ray-induced mutagenesis for enhancing crops extending beyond the Earth's surface. Utilizing gamma rays to improve crop characteristics presents opportunities for agricultural advancement, effectively tackling the worldwide issues of food security and sustainable agriculture.

Chapter 4 explores the interaction between gamma irradiation and entomopathogenic fungi, explicitly investigating how they jointly impact the biological characteristics of *Galleria mellonella* L. This study enhances our comprehension of the intricate interplay between gamma radiation and biological organisms, with potential ramifications for pest management tactics.

Chapter 5 examines the impact of gamma irradiation on the improvement of the nutrition and flavor quality of soybeans, demonstrating the potential of gamma rays in enhancing food quality and nutritional value. This application caters to the increasing demand for sustainable food production solutions and emphasizes the influence of gamma rays on dietary results.

As you begin this journey via this book's pages, you will learn about the various uses of gamma rays, ranging from their application in prospecting and agriculture to their influence on biological systems. Every chapter serves as evidence of the dynamic characteristics of gamma rays and their extensive consequences in several scientific fields.

This book aims to ignite your interest and serve as a helpful resource for researchers, students, and enthusiasts keen to delve into the diverse applications of gamma rays in science and technology.

The editors would like to thank all the contributors for sharing their expertise and Publishing Process Manager Ms. Mirna Papuga at IntechOpen for her continuous assistance in finalizing this work. We are grateful to the publishers for their support in bringing out this timely volume. We hope the book will serve as a valuable resource for students, researchers, and professionals interested in understanding the multifaceted applications of gamma rays.

Hosam M. Saleh and Amal I. Hassan
Egyptian Atomic Energy Authority,
Cairo, Egypt

Section 1

Applications of Gamma Rays

Chapter 1

Introductory Chapter: Exploring the Multifaceted Applications of Gamma Rays in Science and Technology

Hosam M. Saleh and Amal I. Hassan

1. Introduction

The physics of gamma rays, which occupy the highest energy end of the electromagnetic spectrum, is intriguing and complex. Although gamma rays are most often linked with the most powerful and explosive phenomena in the cosmos, they also have significant consequences and uses on our home planet. This introductory chapter tries to demystify gamma rays, explaining their existence, origin, and the countless ways in which they impact and contribute to different scientific areas.

The gamma ray is the highest frequency and most powerful kind of electromagnetic energy. Gamma rays are electromagnetic radiation with a frequency of more than 10 exahertz and a wavelength shorter than 10 picometers, making them distinct from other forms of electromagnetic radiation [1].

Gamma rays' origins are just as interesting as their characteristics. The decay of radioactive nuclei, also known as gamma decay, is a major contributor to this kind of radiation because it occurs when an unstable atomic nucleus releases a gamma ray as a means of dissipating its excess energy. This phenomenon is fundamental to comprehending many radioactive processes occurring on our planet [2, 3].

Also, gamma rays may come from sources other than Earth. These very powerful rays may be found in abundance throughout the universe. The ubiquitous gamma-ray background is mostly fueled by violent cosmic phenomena, including supernovae, the explosive death of big stars, and the interactions surrounding black holes and neutron star mergers. These astronomical occurrences reveal the tremendous power and rapid processes taking place in the depths of space [4].

The unique ability of gamma rays to interact with matter sets them different from other electromagnetic waves. Their enormous intensity enables them to penetrate materials more deeply than X-rays, making them a weapon of great usefulness and peril. This penetrating strength is used in radiation treatment for cancer, which destroys cancer cells [5]. However, the ionizing nature of gamma rays, which may disrupt chemical bonds and cause biological damage, presents considerable hurdles in terms of safety and radiation protection [5].

Scintillation detectors, semiconductor detectors, and Geiger-Müller tubes are only some of the specialized pieces of kit needed to detect and measure gamma rays. Due to the tremendous penetrating force and energy of gamma rays, these instruments are essential to their investigation.

Overall, gamma rays constitute a noteworthy and strong component of the electromagnetic spectrum [6]. Their unusual penetrating force and high energy, together with their generation from both terrestrial and cosmic processes, set them apart from other kinds of electromagnetic radiation, and have made them a topic of continuing interest and research in the domains of physics, astronomy, and medicine.

2. Applications and benefits

The ongoing investigation of gamma rays has been essential in uncovering novel and captivating prospects inside several scientific domains. An area that shows significant potential is the development of gamma ray lasers, which have the capability to produce gamma radiation beams with great coherence, similar to laser beams in the visible spectrum. These gadgets have the potential to revolutionize several fields such as medical imaging, security scanning, manufacturing, and other relevant applications. Simultaneously, enhanced comprehension of gamma ray interactions facilitates progress in radiation therapy, wherein procedures such as gamma knife surgery contribute to the treatment of previously incurable tumors by precisely targeting the affected tissue [7, 8]. The ongoing investigation of gamma rays has been essential in uncovering novel and captivating prospects inside several scientific domains. An area that shows significant potential is the development of gamma ray lasers, which have the capability to produce gamma radiation beams with great coherence, similar to laser beams in the visible spectrum. These gadgets have the potential to revolutionize several fields such as medical imaging, security scanning, manufacturing, and other relevant applications. Simultaneously, enhanced comprehension of gamma ray interactions facilitates progress in radiation therapy, wherein procedures such as gamma knife surgery contribute to the treatment of previously incurable tumors by precisely targeting the affected tissue.

Positron emission tomography (PET) scans are a prominent use of gamma rays within the field of diagnostic imaging (**Figure 1**). PET scans employ a radioactive tracer, often a type of glucose, which is injected into the patient's bloodstream. The tracer releases positrons that engage in interactions with electrons inside the body, leading to the subsequent emission of gamma rays [9]. Subsequently, the PET scanner captures and detects these rays, generating intricate visual representations of the interior anatomical systems inside the body.

Positron emission tomography (PET) scans are of significant importance in the field of cancer due to their ability to provide insights into the metabolic activity of cells. Cancer cells exhibit enhanced metabolic activity, resulting in increased uptake of the radioactive tracer and thus manifesting as heightened brightness on positron emission tomography (PET) pictures. The inclusion of this characteristic makes PET scans a very effective modality for the detection of cancer, evaluation of tumor metastasis, and surveillance of the efficacy of cancer therapies [10]. The use of gamma rays in medicine brings numerous benefits. In treatment, the precision of techniques like Gamma Knife surgery allows for effective treatment of tumors with reduced risk and discomfort compared to traditional surgery. In diagnostics, PET scans provide detailed images that are crucial for early detection and treatment planning [11].

However, the use of gamma rays is not without risks. Due to their high energy and ionizing nature, gamma rays can cause damage to living tissues. This necessitates

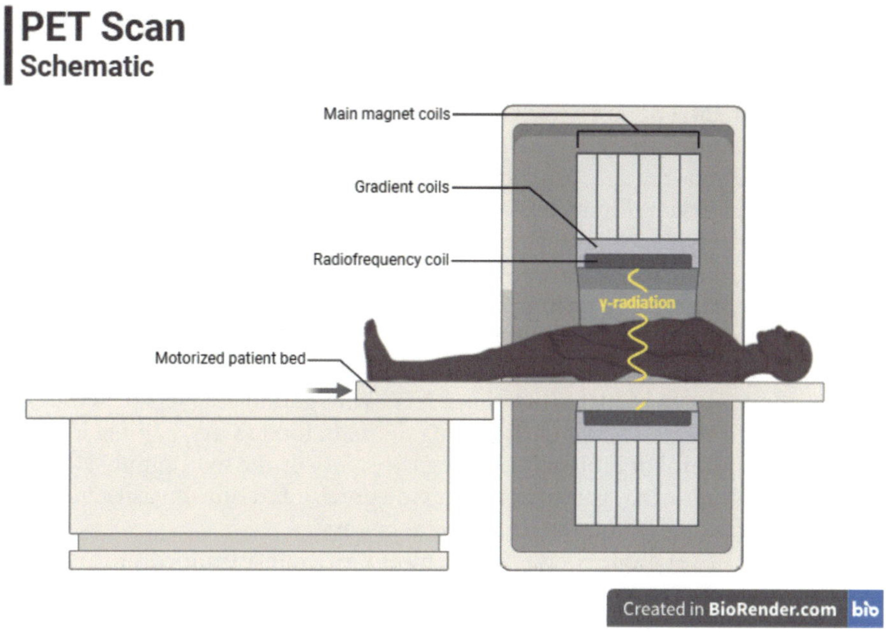

Figure 1.
PET scan schematic.

careful control and minimization of exposure to protect patients and medical staff. The risks are mitigated through precise targeting in treatments and minimal tracer usage in diagnostics, but they remain a critical consideration in the medical use of gamma rays [12].

Gamma rays, with their unique properties, have found a wide range of applications in various industrial and environmental contexts, enhancing safety, efficiency, and quality in these sectors. One of the primary industrial uses of gamma rays is in non-destructive testing (NDT). This process involves inspecting materials, components, or assemblies for defects without causing damage. Gamma rays, due to their high penetrative power, are ideal for this purpose. They can penetrate various materials, including metals, plastics, and ceramics, allowing for the internal examination of objects [13].

In NDT, gamma rays are used to create a radiographic image of the item being inspected. This is similar to taking an X-ray in medical imaging. The gamma rays pass through the object and are captured on a detector or film on the other side [14]. Any flaws in the material, such as cracks, voids, or inclusions, will absorb or block the gamma rays differently than the surrounding material, appearing as distinct anomalies on the radiograph. This method is widely used in various industries, including aerospace, automotive, and construction, to ensure the integrity and safety of components.

Gamma rays are also used in food irradiation, a process to ensure food safety and extend shelf life. Food irradiation involves exposing food to gamma rays to kill bacteria, parasites, and other pathogens, reducing the risk of foodborne illnesses [15]. It also delays the ripening and sprouting in fruits and vegetables, which helps preserve their freshness during transportation and storage.

The process of food irradiation is done using sources like Cobalt-60, which emits gamma rays. The key advantage of using gamma rays is that they effectively sterilize food without raising its temperature, keeping the food fresh and preventing the changes in texture or taste often associated with thermal processing [16].

In environmental monitoring, gamma rays are used to analyze and monitor various environmental parameters. They can be employed in the measurement of soil density and moisture content, which is essential for agriculture and construction projects. Gamma-ray spectrometry is also used to monitor and assess the levels of natural and artificial radionuclides in the environment, an important aspect of environmental safety, especially around nuclear facilities [17].

Furthermore, gamma rays are used in tracking and monitoring pollution. For example, gamma-ray tracing techniques can track the movement of sediments in rivers and oceans, helping in the study of erosion and sedimentation patterns [17].

The use of gamma rays in these applications offers numerous benefits in terms of safety and efficiency. In industrial settings, NDT with gamma rays ensures the safety and reliability of products without damaging them. In food safety, gamma irradiation eliminates harmful organisms, ensuring the safety of the food supply [15]. In environmental monitoring, gamma rays provide precise and non-invasive means of analyzing and tracking various environmental factors.

However, the use of gamma rays also requires strict safety protocols to protect workers and the environment from radiation exposure. Proper shielding, handling, and storage of gamma-ray sources are essential to ensure safety in these applications.

The safe handling and regulation of gamma rays are paramount due to their high energy and potential for biological damage. Understanding and adhering to safety measures and regulatory frameworks are essential to mitigate risks associated with gamma radiation in various applications, from medical treatments to industrial processes [18].

One of the primary safety measures in working with gamma rays is the use of adequate shielding. Materials such as lead, concrete, or specialized alloys are often used to absorb or reduce gamma radiation, thereby protecting personnel and the environment [19].

Minimizing the time spent near gamma radiation sources and maximizing the distance from them are key safety practices. This reduces the exposure and, consequently, the dose of radiation received. In environments where gamma radiation is present, wearing appropriate protective equipment, such as lead aprons or shields, is crucial for minimizing exposure. The use of radiation detectors and dosimeters is essential for monitoring radiation levels and individual exposure. These devices help in ensuring that exposure stays within safe limits [20].

Comprehensive training for personnel working with gamma rays is essential. This includes understanding the properties of gamma radiation, knowing the safety protocols, and being prepared for emergency situations.

The International Atomic Energy Agency (IAEA) and the World Health Organization (WHO) play a pivotal role in outlining guidelines and standards for the safe utilization of gamma rays, especially in medical and industrial contexts. Countries typically have their own regulatory bodies and specific regulations to govern the use of radiation. These national regulations dictate permissible exposure limits, safety protocols, and requirements for obtaining licenses and conducting inspections [21]. Entities employing gamma rays are mandated to acquire the necessary licenses. They are also subject to routine inspections to ensure adherence to safety standards. Additionally, regulations encompass the safe management, storage, and disposal of radioactive materials and waste, which is vital for averting environmental contamination and public exposure.

The responsible handling of gamma rays transcends regulatory compliance; it is a moral obligation to guarantee public and environmental safety. Improper management or negligence in the handling of gamma rays can result in severe health issues like radiation sickness, heightened cancer risks, and environmental damage [22].

In medical settings, gamma rays are indispensable for diagnostics and treatment. However, their application necessitates a delicate equilibrium between therapeutic advantages and potential risks. Practices such as precise targeting in Gamma Knife surgery and minimizing unnecessary exposure in diagnostic processes are crucial [23]. In industrial and environmental sectors, safety protocols are crucial to safeguard workers and the public from unintended exposure. This is especially significant in industries like aerospace and construction, where material integrity is essential for safety.

3. Environmental monitoring

Environmental monitoring plays a crucial role in comprehending and controlling the well-being of our ecosystems. Gamma rays have become a potent means of evaluating and tracking environmental conditions, out of the many tools and approaches used [24]. Gamma rays are a form of electromagnetic radiation with high energy that is emitted from the nuclei of atoms. They possess the ability to permeate through a majority of substances, rendering them valuable in the identification and quantification of diverse chemicals, particularly those that are radioactive [25].

Gamma-ray spectroscopy is employed to cartographically identify both naturally occurring and human-made radioactive substances present in the surroundings. Gamma-ray spectroscopy exploits the phenomenon that distinct radioactive isotopes release distinctive gamma rays at precise energy levels. Through the utilization of detectors capable of correctly quantifying the exact energy of gamma rays released from a specific place, scientists can determine the types and quantities of radioactive elements present [26].

Gamma spectroscopy can be employed in aerial radiation mapping studies to detect and identify both naturally occurring and artificially generated radioisotopes in the environment. Helicopters and airplanes fitted with gamma spectrometers conduct aerial surveys in a systematic pattern over a region, collecting data that is combined with GPS information to generate intricate maps illustrating radiation levels throughout the terrain [27]. These maps have the ability to identify higher levels of uranium, thorium, potassium-40, or other radioactive elements that are linked to specific rocks or minerals.

Gamma spectroscopic mapping allows for the detection and monitoring of man-made radioactive pollution resulting from nuclear accidents, fallout from weapons testing, and incorrect disposal of radioactive waste, among other sources. The Chernobyl tragedy, for example, enabled scientists to comprehensively determine the extent to which radiation spread. Following the Fukushima nuclear accident, highly sensitive equipment was used to identify even small amounts of pollution on the opposite side of the Pacific Ocean [28].

Gamma spectroscopy mapping data provides accurate radiological profiles of landscapes and exact distribution patterns of isotopes. This information enables authorities to enhance their monitoring of environmental radioactivity, evaluate hazards, respond effectively to emergencies, find misplaced radioactive sources, and do other related tasks.

Gamma-ray equipment is employed to monitor post-nuclear incident environments and evaluate the dispersion and density of radioactive pollutants. Following an accidental or intentional nuclear release, it is crucial to promptly assess the extent and scale of radioactive contamination in the affected regions. Utilizing gamma-ray surveying instruments on vehicles, drones, robots, or handled by reaction workers can be highly advantageous for this undertaking [29].

Cesium-137, a type of fission product, emits gamma rays that have the ability to permeate through several commonly found materials. Gamma detectors with high sensitivity and precise directionality, along with global positioning systems, have the capability to monitor contamination levels in real-time. This enables the identification and mapping of dangers with precise geographical accuracy, which may then be used to inform decisions on evacuation and remediation [30]. Utilizing aerial gamma surveys conducted by manned helicopters or drones provides responders with rapid acquisition of high-resolution data, especially in cases where infrastructure has been significantly compromised.

Both air sampling using particulate filters and ground-level gamma scans are essential for comprehensive environmental monitoring. Isotope identities provide confirmation of the specific nuclear materials that were released, which is crucial for supporting the medical treatment of those who have been exposed if such treatment is available. Conducting multiple surveys enables the measurement of changes in radioactive decay and contaminant movement over a period of time [31].

State-of-the-art spectroscopic gamma detectors offer the highest level of accuracy in measurements. When used in conjunction with suitable modeling approaches, this technology has the capability to produce 3D contamination maps that provide estimates of material depth and activity levels [32]. Ensuring high-quality monitoring after an incident is crucial for effectively controlling health risks and providing guidance for intricate, prolonged remediation processes.

Gamma rays are used to ascertain the concentration of natural radionuclides in soils, a vital factor in comprehending soil health and fertility. The concentration of natural radioisotopes in soil is directly associated with important indices of soil quality, such as mineral content, texture, and cation exchange capacity (CEC) - which refers to the soil's ability to retain nutrients [33]. Gamma-ray spectroscopy enables the precise quantification of radioisotopes such as potassium-40, uranium-238, thorium-232, and radioactive cesium, which are commonly found in clay and organic materials.

Utilizing airborne gamma-ray sensors to survey extensive agricultural regions can effectively delineate the fluctuation in these radioisotopes. By utilizing this method, it becomes possible to draw conclusions regarding the distribution of soil composition zones and the variations in fertility levels throughout extensive territories [34]. This aids in the identification of regions with lower nutrient levels or water retention capacities, therefore providing valuable insights for agricultural management strategies.

Portable gamma-ray spectrometers that are located on the ground can also investigate the concentrations of radioisotopes. The examination of paired soil samples in a laboratory improves the understanding of the relationships between radioisotope signals, underlying geology, and agricultural qualities such as permeability, porosity, drainage class, and particle size distribution [34]. This data enhances comprehension of soil quality at smaller, more specific scales.

Gamma rays possess non-invasive properties and exceptional penetration capabilities, rendering them an exceptionally valuable tool for deducing both physical and chemical attributes of soil. Gamma scanning, when used in soil genesis studies, can reveal regional variations in weathering and geochemical history [35]. These insights can assist in directing the process of restoring soils that have been damaged.

The utilization of gamma-ray monitoring is implemented to identify and measure radionuclides in bodies of water, guaranteeing the security of potable water and aquatic ecosystems. In response to the Fukushima nuclear accident, authorities in the United States and Canada established coastal monitoring programs utilizing gamma spectrometry to monitor the dispersion of cesium, iodine, and other radioactive isotopes carried over the Pacific Ocean and into coastal waterways [36]. This facilitated the prompt identification of any potential contamination hazards. Gamma dose rate meters installed on floating devices and buoys, which are linked to wireless networks, have the capability to offer immediate and continuous monitoring of radiation levels in reservoir water. They can even detect sudden releases of radiation from nuclear reactors [37].

Drinking water quality can be affected by the presence of natural radioisotopes in aquifer rock and sediments. In a study conducted in 2022, gamma spectroscopy was employed to examine the levels of radium in water distribution infrastructure. This analysis helped to identify specific places and mechanisms that contribute to higher levels of radioactivity in consumer tap water [38]. Researchers are currently working on the development of portable gamma detectors for quick and cost-effective monitoring of water produced from fracking operations. These detectors will be used with lab-based ICP-MS analysis to detect short half-life radionuclides.

Gamma-ray analysis has the capability to identify and measure radionuclides present in the air, which serve as reliable markers of both pollution and nuclear activity. Natural radionuclides such as radon-222, as well as manufactured isotopes, have the potential to accumulate in the surrounding air, which can be harmful if inhaled. According to Shakhashiro et al. [39], gamma spectroscopy is the most sensitive method for detecting and determining the exact composition of radioactive substances in the air.

Networks of air particle monitoring stations equipped with gamma detectors provide uninterrupted data on ambient radiation levels and radioactive concentrations. It is necessary to investigate any abnormal increases in order to identify potential sources of pollution or radioactive releases before the general population is exposed to them [40]. The ability to be aware of events as they happen in real time enables authorities to make timely and effective decisions.

Various incident scenarios can result in the release of gamma-emitting radioisotopes into the atmosphere. These encompass incidents at nuclear sites, mishandling of medicinal or industrial isotopes, detonation of radiological dispersal devices, and the dissemination of fallout particles that have been stirred up again. Identifying distinct fission product signals aids in determining the origin [41].

Drones and unmanned robots, which are equipped with small gamma spectrometers, provide flexibility in mapping incidents from the air and ground, while also lowering the amount of radiation exposure for responders [42]. Understanding the dispersal of pollution is crucial for both protection and repair efforts.

Scientists employ gamma rays to investigate the effects of radiation on flora and fauna, thereby aiding in the preservation of biodiversity. Living beings can be affected by gamma rays through many methods, which result in the transfer of energy that can modify cell function and DNA. The process of quantifying radiation effects aids in the prediction of ecosystem stability and the establishment of safety limits for animals concerning contamination events or prolonged exposure [43].

Field gamma spectrometry surveys assess the radiation levels in both natural background and elevated environments by mapping the distribution of radionuclides in soil and food chains. The study conducted by Nikolova et al. [44] demonstrates the use of paired biological sampling to investigate the dynamics of radioisotope transfer and to establish benchmarks for biodiversity.

The study conducted by Thiemann et al. [45] involves analyzing the genomes, physiology, and morphology of animals from multiple generations that have been exposed to controlled gamma, x-ray, or ion beam irradiation. The purpose of this study is to understand the rates of mutations and the mechanisms of adaption. This demonstrates the diversity in resistance to radiation among different species.

Some plant studies have shown that exposure to radiation can increase genetic variety, which may help protect ecosystems against environmental changes at low doses [46]. Nevertheless, some studies have observed reduced germination, slowed growth, and lethal effects beyond specific dosage thresholds [47].

Gamma-ray spectroscopy is an advanced analytical method employed to detect and measure radionuclides in samples from the environment. Through the examination of gamma-ray spectra, which represent the dispersion of gamma-ray energy released by radioactive substances, this technique may precisely ascertain the specific types and quantities of radionuclides that are present [48]. Mastering this method is crucial for comprehending ambient radioactivity, particularly in regions impacted by nuclear operations or mishaps. The method's precision and dependability establish it as a fundamental technique in the monitoring of environmental radiation. Accurate interpretation and analysis of gamma-ray spectra necessitate a specialist understanding of nuclear physics due to their intricate complexity.

Portable gamma-ray detectors are essential in field studies. They enable the collection of measurements directly at the location and prompt examination of environmental samples [49]. These instruments are highly valuable for quickly evaluating radiation levels in many contexts, ranging from natural landscapes to metropolitan surroundings. Their portability allows researchers and workers to carry out surveys in regions where laboratory analysis may be impractical or unfeasible. Although these detectors are highly adaptable, they frequently encounter constraints in terms of sensitivity and accuracy when compared to more advanced laboratory equipment [49].

Satellite and aerial-based gamma-ray sensors enable remote sensing, providing the ability to map broad areas for environmental monitoring purposes. This technology is crucial for conducting large-scale surveys, enabling the identification and cartography of radioactive substances across wide areas [50]. It is highly efficient in locations that are difficult to reach or pose risks, as it can gather crucial data without requiring the collection of physical samples. These technologies play a crucial role in conducting extensive environmental evaluations, particularly in monitoring the spread of radioactive pollutants after a nuclear incident [51].

Although these technologies provide substantial advantages, their deployment involves several problems and concerns. Deciphering gamma-ray data is intricate and necessitates a profound comprehension of nuclear physics and environmental science. The complexity of the data necessitates careful analysis, as any misinterpretation can result in erroneous conclusions regarding environmental conditions. The sensitivity and precision of gamma-ray devices, particularly portable detectors, can occasionally restrict the range of detection and analysis. This constraint is a pivotal element for carrying out intricate investigations or handling radiation at a minimal intensity [29]. Strict compliance with safety regulations and regulatory criteria is necessary while using gamma-ray equipment. Comprehensive safety procedures are required while handling radioactive materials and operating equipment that emits gamma rays in order to ensure the protection of both operators and the environment.

4. Expanding applications and future potential

Gamma rays possess distinctive properties including penetration, sterilization, and material activation capabilities that researchers continue finding innovative applications for across scientific and industrial fields.

Gamma radiography gauges enable non-destructive analysis of density variances and physical integrity in materials from concrete to aerospace composites [52].

Food and polymer irradiation harnesses high gamma energy to alter chemical or molecular structures, creating improved material properties or killing pathogens [53].

Emerging x-ray fluorescence and gamma-ray computed tomography scans paired with carbon dating verify the age, origins, and preservation states of rare organic items [54].

Gamma-ray scanning portals provide drive-through screening of shipping containers, catching shielded nuclear materials [55]. Detection networks leverage signals from fission events for tracking weapons tests and detonations with high accuracy. 3D gamma spectroscopy mapping of nuclear forensics debris pinpoints device attributes aiding attribution [56]. Lab demonstrations of modulated gamma-ray optical communication signals exhibit resistance to scattering with potential for ultra-secure space transmissions [57]. The versatility and value proposition of gamma radiation will continue to extend across various applications in the future decades as advancements in source, sensor, compute, and data fusion technologies evolve [58]. The immense energy of gamma rays could potentially be utilized in the realm of quantum computing and communication. While this use is more speculative, the possibility of using gamma rays to manipulate quantum states has the potential to create new opportunities in this advanced subject [58]. Investigating the capacity of gamma rays for energy generation, specifically in domains such as nuclear fusion, holds promise as a prospective field of study. Although it presents difficulties, utilizing gamma rays for energy has the potential to offer a novel approach to clean and efficient power production [59]. With the progression of technology, the possibilities for using gamma rays are expected to broaden, presenting inventive solutions and unveiling novel opportunities in the fields of research and technology.

Author details

Hosam M. Saleh* and Amal I. Hassan
Radioisotope Department, Nuclear Research Center, Egyptian Atomic Energy Authority, Egypt

*Address all correspondence to: hosamsaleh70@yahoo.com

References

[1] Philippov A, Kramer M. Pulsar magnetospheres and their radiation. Annual Review of Astronomy and Astrophysics. 2022;**60**:495-558

[2] Gupta TK, Gupta TK. Nuclear radiation, ionization, and radioactivity. In: Radiation, Ionization, and Detection in Nuclear Medicine. United States: Springer; 2013. pp. 1-57

[3] Vallabhajosula S. Radioactivity. In: Molecular imaging and targeted therapy: Radiopharmaceuticals and clinical applications. United States: Springer; 2023. pp. 49-62

[4] Diehl R, Korn AJ, Leibundgut B, Lugaro M, Wallner A. Cosmic nucleosynthesis: A multi-messenger challenge. Progress in Particle and Nuclear Physics. Vol. 127. Amsterdam, Netherlands: Elsevier; 2022. pp. 103983

[5] Persson BRR. Radiation hazards. In: Occupational Hazards in the Health Professions. CRC Press; 2020. pp. 163-235

[6] Giommi P, Padovani P, Polenta G. A simplified view of blazars: The γ-ray case. Monthly Notices of the Royal Astronomical Society. Oxford University. 2013;**431**(2):1914-1922

[7] Lindquist C. Gamma knife radiosurgery. In: Seminars in Radiation Oncology. Amsterdam: Elsevier; 1995. pp. 197-202

[8] Niranjan A, Bowden G, Flickinger JC, Lunsford LD. Gamma knife radiosurgery. Principles and Practice of Stereotactic Radiosurgery. United States: Springer; 2015. pp. 111-119

[9] Terziċ A, Goerres G. Diagnostic imaging–positron emission tomography, combined PET/CT imaging. In: Osteomyelitis of the Jaws. United States: Springer; 2008. pp. 113-119

[10] Gallamini A, Zwarthoed C, Borra A. Positron emission tomography (PET) in oncology. Cancers (Basel). 2014;**6**:1821-1889

[11] Reynaert N. PET and MRI based RT treatment planning: Handling uncertainties. Cancer/Radiothérapie. 2019;**23**:753-760

[12] Harrell CR, Djonov V, Fellabaum C, Volarevic V. Risks of using sterilization by gamma radiation: The other side of the coin. International Journal of Medical Sciences. 2018;**15**:274

[13] Wang B, Zhong S, Lee T-L, Fancey KS, Mi J. Non-destructive testing and evaluation of composite materials/structures: A state-of-the-art review. Advances in Mechanical Engineering. 2020;**12**:1687814020913761

[14] Rapaport MS, Gayer A. Application of gamma ray computed tomography to nondestructive testing. NDT E International. 1991;**24**:141-144

[15] Indiarto R, Pratama AW, Sari TI, Theodora HC. Food irradiation technology: A review of the uses and their capabilities. International Journal of Engineering Trends and Technology (IJETT). 2020;**68**:91-98

[16] Ashraf S, Sood M, Bandral JD, Trilokia M, Manzoor M. Food irradiation: A review. International Journal of Chemical Studies. 2019;**7**:131-136

[17] Goldsten JO, Rhodes EA, Boynton WV, Feldman WC, Lawrence DJ, Trombka JI, et al. The MESSENGER gamma-ray and neutron spectrometer.

The Messenger Mission to Mercury. United States: Springer; 2007. pp. 339-391

[18] Sarkar PS. Safety, regulations, metrology and standards in neutron imaging. In: Neutron Imaging: Basics, Techniques and Applications. Germany: Springer; 2022. pp. 207-235

[19] Kaur T, Sharma J, Singh T. Review on scope of metallic alloys in gamma rays shield designing. Progress in Nuclear Energy. 2019;**113**:95-113

[20] NG SE, SA F. Assessment of awareness and practice of ionizing radiation protection procedures among exposed health care workers. Egyptian Journal of Occupational Medicine. 2020;**44**:529-544

[21] Anurag L, Gaurav M, Sanjay D, Chandrashekar M. Spectrum of IAEA standard with regard to radiation application. Journal of Datta Meghe Institute of Medical Sciences University. 2022;**17**:461-467

[22] Avendaño G. Critical importance of multilateral studies related with adverse events in medical devices. Health and Technology (Berl). 2016;**6**:213-227

[23] Luharia A, Mishra G, Saroj D, Sonwani V, Dhoble SJ. The role of physics in modern radiotherapy: Current advances and developments. Photophysics and Nanophysics in Therapeutics. USA: Elsevier; 2022. pp. 139-162

[24] Rodrigues EA, Ferreira ML, de Carvalho AR, Bustillos JOWV, Victor RABM, Sodré MG, et al. Land, water, and climate issues in large and megacities under the lens of nuclear science: An approach for achieving sustainable development goal (SDG11). Sustainability. 2022;**14**:13646

[25] Mosayebnia M, Ahmadi M, Emzhik M, Hajiramezanali M. Gamma-ray involved in cancer therapy and imaging. In: Electromagnetic Waves-Based Cancer Diagnosis and Therapy. Amsterdam, Netherlands: Elsevier; 2023. pp. 295-345

[26] El-Taher A. Nuclear analytical techniques for detection of rare earth elements. International Journal of Radiation Applications and Instrumentation. 2018;**3**:53-64

[27] Varley A, Tyler A, Smith L, Dale P, Davies M. Mapping the spatial distribution and activity of 226Ra at legacy sites through Machine Learning interpretation of gamma-ray spectrometry data. Science of the Total Environment. 2016;**545**:654-661

[28] Buesseler K, Dai M, Aoyama M, Benitez-Nelson C, Charmasson S, Higley K, et al. Fukushima Daiichi–derived radionuclides in the ocean: transport, fate, and impacts. Annual Review of Marine Science. 2017;**9**:173-203

[29] Marques L, Vale A, Vaz P. State-of-the-art mobile radiation detection systems for different scenarios. Sensors. 2021;**21**:1051

[30] Randall GL, Iglesis E, Wong HF, Speller RS. A method of providing directionality for ionising radiation detectors—RadICAL. Journal of Instrumentation. 2014;**9**:P10011

[31] Chartin C, Evrard O, Onda Y, Patin J, Lefèvre I, Ottlé C, et al. Tracking the early dispersion of contaminated sediment along rivers draining the Fukushima radioactive pollution plume. Anthropocene. 2013;**1**:23-34

[32] Verbelen Y, Martin PG, Ahmad K, Kaluvan S, Scott TB. Miniaturised low-cost gamma scanning platform for contamination identification, localisation

and characterisation: A new instrument in the decommissioning toolkit. Sensors. 2021;**21**:2884

[33] Silvero NEQ, Demattê JAM, Minasny B, Rosin NA, Nascimento JG, Albarracín HSR, et al. Sensing technologies for characterizing and monitoring soil functions: A review. Advances in Agronomy. 2023;**177**:125-168

[34] Moonjun R, Shrestha DP, Jetten VG, van Ruitenbeek FJA. Application of airborne gamma-ray imagery to assist soil survey: A case study from Thailand. Geoderma. 2017;**289**:196-212

[35] Guagliardi I, Rovella N, Apollaro C, Bloise A, De Rosa R, Scarciglia F, et al. Effects of source rocks, soil features and climate on natural gamma radioactivity in the Crati valley (Calabria, Southern Italy). Chemosphere. 2016;**150**:97-108

[36] Cooke MW, Trudel M, Gurney-Smith HJ, Kellogg JP, Cullen JT, Francisco BBA, et al. Radioactivity concentration measurements in fish and shellfish samples from the west coast of Canada after the Fukushima nuclear accident (2011-2018). Journal of Environmental Radioactivity. 2022;**251**:106934

[37] Arndt J, Kirchner JS, Jewell KS, Schluesener MP, Wick A, Ternes TA, et al. Making waves: Time for chemical surface water quality monitoring to catch up with its technical potential. Water Research. 2022;**213**:118168

[38] Vengosh A, Coyte RM, Podgorski J, Johnson TM. A critical review on the occurrence and distribution of the uranium-and thorium-decay nuclides and their effect on the quality of groundwater. Science of the Total Environment. 2022;**808**:151914

[39] Shakhashiro A, Doherty P, Logar JK, Vodenik B, Verheyen L, Taggart M. New certified reference materials and proficiency test for environmental radioactivity measurements. Accreditation and Quality Assurance. 2016;**21**:351-360

[40] Barescut JC, Gariel JC, Péres JM, Balonov M, Linsley G, Louvat D, et al. The IAEA standards for the radioactive discharge control: Present status and future development. Radioprotection. 2005;**40**:S721-S726

[41] Vilardi I, Antonacci G, Battisti P, Castellani CM, Ciciani L, Del Gaudio D, et al. Large-scale individual monitoring of internal contamination by gamma-emitting radionuclides in nuclear accident scenarios. Journal of Radiological Protection. 2019;**40**:134

[42] Martin PG, Kwong S, Smith NT, Yamashiki Y, Payton OD, Russell-Pavier FS, et al. 3D unmanned aerial vehicle radiation mapping for assessing contaminant distribution and mobility. International Journal of Applied Earth Observation and Geoinformation. Vol. 52. Amsterdam, Netherlands: Elsevier; 2016. pp. 12-19

[43] Agency IAE. Effects of ionizing radiation on plants and animals at levels implied by current radiation protection standards. IAEA; 1992

[44] Nikolova I, Johanson KJ, Clegg S. The accumulation of 137Cs in the biological compartment of forest soils. Journal of Environmental Radioactivity. 2000;**47**:319-326

[45] Thiemann RA, Thornton JK, Stayer PA, Riley E, Clark R, Armour N, et al. A Novel presentation of clostridium perfringens in young broilers. Avian Diseases. 2022;**66**:205-212

[46] Maity TN, Saha AK, Dubey A, Laha R. Search for dark matter using

sub-PeV γ-rays observed by Tibet AS γ. Physical Review D. 2022;**105**:L041301

[47] Tran TLC, Callahan DL, Islam MT, Wang Y, Arioli T, Cahill D. Comparative metabolomic profiling of Arabidopsis thaliana roots and leaves reveals complex response mechanisms induced by a seaweed extract. Frontiers in Plant Science. 2023;**14**:1114172

[48] Mahmood HS, Hoogmoed WB, Van Henten EJ. Proximal gamma-ray spectroscopy to predict soil properties using windows and full-spectrum analysis methods. Sensors. 2013;**13**:16263-16280

[49] Al Hamrashdi H, Cheneler D, Monk SD. A fast and portable imager for neutron and gamma emitting radionuclides. Nuclear Instruments and Methods in Physics Research Section A: Accelerators, Spectrometers, Detectors and Associated Equipment. 2020;**953**:163253

[50] Zhang J, Huang Y, Pu R, Gonzalez-Moreno P, Yuan L, Wu K, et al. Monitoring plant diseases and pests through remote sensing technology: A review. Computers and Electronics in Agriculture. 2019;**165**:104943

[51] Abdulraheem MI, Zhang W, Li S, Moshayedi AJ, Farooque AA, Hu J. Advancement of remote sensing for soil measurements and applications: A comprehensive review. Sustainability. 2023;**15**:15444

[52] Groysman A. Nondestructive testing and corrosion monitoring. Non-Destructive Evaluation of Corrosion and Corrosion-assisted Cracking. Hoboken, New Jersey, U.S: Wiley; 2019. pp. 261-409

[53] Farkas J, Mohácsi-Farkas C. History and future of food irradiation. Trends in Food Science and Technology. 2011;**22**:121-126

[54] Miyahara AAL, Paixão CP, dos Santos DR, Pagin-Cláudio F, da Silva GJ, Bertoleti IAF, et al. Urban dendrochronology toolkit for evidence-based decision-making on climate risk, cultural heritage, environmental pollution, and tree management–A systematic review. Environmental Science & Policy. 2022;**137**:152-163

[55] Shakhashiro A, Sansone U, Wershofen H, Bollhöfer A, Kim CK, Kim CS, et al. The new IAEA reference material: IAEA-434 technologically enhanced naturally occurring radioactive materials (TENORM) in phosphogypsum. Applied Radiation and Isotopes. 2011;**69**:231-236

[56] Varga Z, Wallenius M, Krachler M, Rauff-Nisthar N, Fongaro L, Knott A, et al. Trends and perspectives in nuclear forensic science. TrAC Trends in Analytical Chemistry. 2022;**146**:116503

[57] De Angelis A, Tatischeff V, Argan A, Brandt S, Bulgarelli A, Bykov A, et al. Gamma-ray astrophysics in the MeV range: The ASTROGAM concept and beyond. Experimental Astronomy. 2021;**51**:1225-1254

[58] Wright T, West A, Licata M, Hawes N, Lennox B. Simulating ionising radiation in gazebo for robotic nuclear inspection challenges. Robotics. 2021;**10**:86

[59] Howell CR, Ahmed MW, Afanasev A, Alesini D, Annand JRM, Aprahamian A, et al. International workshop on next generation gamma-ray source. Journal of Physics. G, Nuclear and Particle Physics: An Institute of Physics Journal. 2021;**49**:10502

Chapter 2

Gamma Ray Induced Mutagenesis for Crop Improvement: Applications, Advancements, and Challenges

Godswill Ntsomboh Ntsefong, Fokam Paul Ernest, Likeng-Li-Ngue Benoit Constant, Tabi Mbi Kinsley, Zambou Alain Hervé, Mafouasson Hortense Noelle and Bell Joseph Martin

Abstract

Gamma ray induced mutagenesis is a powerful tool for crop improvement that has been used for decades to generate genetic variability in crops. This method has advantages over other mutagenic agents due to its high penetrance and ability to induce a large number of mutations in a single treatment. Recent advancements in high-throughput screening techniques and molecular marker analysis have facilitated the identification and characterization of beneficial traits resulting from gamma ray induced mutagenesis. However, there are also challenges associated with this method, such as the need to balance trait improvements with potential negative effects on crop yield or quality, ethical considerations, safety measures, and considerations for climate-smart agriculture. This chapter provides an overview of the historical background and principles of gamma ray induced mutagenesis, its applications in crop improvement and climate-smart agriculture, recent advancements, challenges, and future directions. The chapter highlights the potential of gamma ray induced mutagenesis for generating new genetic variation in crops and its potential role in addressing global food security and climate change challenges.

Keywords: gamma rays, mutagenesis, crop improvement, climate-smart agriculture, high-throughput screening

1. Introduction

Gamma ray induced mutagenesis, a technique that has been utilized for decades, plays a significant role in crop improvement by introducing genetic variability in crops [1, 2]. By subjecting crops to gamma rays, mutations are induced in their DNA, resulting in the generation of novel genetic variants with desirable traits like increased yield, disease resistance, and improved quality [3]. This method has several advantages over other mutagenic agents, including its high penetrance and ability to induce a large number of mutations in a single treatment [4].

The use of gamma ray induced mutagenesis for crop improvement dates back to the 1920s and has been successfully employed in various crops such as rice, wheat, barley, and soybean [3]. In recent years, its significance has grown due to its potential to address global food security challenges by creating new genetic variations that enhance crop yields and resilience to environmental stresses [2]. As the world population continues to increase, ensuring an adequate and sustainable food supply becomes crucial, making gamma ray induced mutagenesis an invaluable tool for crop breeders and researchers.

Moreover, gamma ray induced mutagenesis intersects with the concept of climate-smart agriculture, which aims to achieve food security, climate change resilience, and sustainable agricultural practices [5]. By generating novel genetic variants with desirable traits such as drought tolerance, this technique contributes to promoting climate-smart agriculture and supports global efforts towards sustainable agriculture [2]. With the intensifying impacts of climate change, including more frequent and severe droughts, it is imperative to develop crops with increased resilience to ensure food production in such challenging environments.

Gamma ray induced mutagenesis offers an effective means to introduce genetic variability and select for desired traits in crops. The induced mutations can result in novel characteristics that might not be present in the existing gene pool, providing breeders with a broader range of options for crop improvement [1]. This approach allows for targeted modification of specific traits without introducing genes from unrelated species, addressing concerns related to genetically modified organisms (GMOs) and promoting public acceptance of crop improvement methods [4].

Despite its advantages, gamma ray induced mutagenesis does have limitations. The process is random, and not all induced mutations lead to desirable outcomes. Additionally, identifying and selecting the desired traits from a large pool of mutants can be time-consuming and labor-intensive [1]. However, advancements in molecular techniques, such as high-throughput genotyping and phenotyping, have facilitated the identification and characterization of mutants with desired traits, expediting the breeding process [2].

Overall, gamma ray induced mutagenesis is a valuable tool for crop improvement, enabling the generation of genetic variability and the selection of desirable traits in crops. Its long history of successful application in various crops and its potential to address global food security challenges make it a vital technique for ensuring sustainable agriculture. Moreover, by contributing to climate-smart agriculture through the creation of genetic variants with traits like drought tolerance, gamma ray induced mutagenesis aligns with the goals of sustainable and resilient agricultural practices. As technology and understanding continue to advance, gamma ray induced mutagenesis holds promise for further enhancing crop productivity and addressing the evolving challenges of food security and climate change.

2. Applications of gamma ray induced mutagenesis in crop improvement

2.1 Advantages of gamma ray induced mutagenesis over other mutagenic agents

Gamma ray induced mutagenesis stands out among other mutagenic agents, including chemical mutagens and transposable elements, due to its numerous advantages. One notable advantage is its ability to induce a high frequency of mutations in a single treatment, thereby facilitating the creation of novel genetic variants with desirable traits [2]. This high mutagenic efficiency is crucial for crop improvement efforts as it provides breeders with a greater pool of genetic diversity to select from.

Another advantage of gamma ray induced mutagenesis is the high penetrance of gamma rays, which allows them to reach and induce mutations in both coding and non-coding regions of the DNA. This broad mutational spectrum enables the generation of diverse mutations that can impact various aspects of crop traits [4]. By targeting both coding and non-coding regions, gamma ray induced mutagenesis increases the likelihood of generating genetic variations that can lead to phenotypic changes and desired agronomic traits.

Compared to chemical mutagens, gamma ray induced mutagenesis offers several benefits. Chemical mutagens often have limited specificity and can induce a wide range of mutations, including undesirable ones. On the other hand, gamma rays provide a more controlled mutagenesis process, allowing breeders to focus on specific traits of interest [2]. This targeted approach reduces the likelihood of introducing unwanted mutations and streamlines the breeding process by narrowing down the selection pool.

Transposable elements, or jumping genes, have been widely used as mutagenic agents in crop improvement. However, their application can be challenging due to their unpredictable behavior and potential to disrupt essential genes. In contrast, gamma rays offer a more predictable and controllable mutagenesis process, making them a preferred choice for many crop breeding programs [4]. The ability to induce mutations in a wide range of genes without the risk of detrimental disruptions provides a valuable advantage in crop improvement endeavors.

The advantages of gamma ray induced mutagenesis extend beyond its technical aspects. This approach has a long history of successful application in crop improvement, with numerous studies demonstrating its effectiveness in generating new genetic variations with desirable traits [2]. The accumulated knowledge and experience gained over the years have contributed to the refinement of protocols and breeding strategies, further enhancing the efficiency of gamma ray induced mutagenesis.

Furthermore, gamma ray induced mutagenesis aligns with the principles of sustainable agriculture and environmental stewardship. The technique does not involve the introduction of genes from unrelated species, alleviating concerns surrounding genetically modified organisms (GMOs) and promoting public acceptance of crop improvement methods [4]. By utilizing the natural properties of gamma rays to induce mutations, breeders can develop crops with improved traits while adhering to sustainable agricultural practices.

In conclusion, gamma ray induced mutagenesis offers several advantages over other mutagenic agents in crop improvement. Its ability to induce a high frequency of mutations and target both coding and non-coding regions of the DNA provides breeders with a diverse pool of genetic variations to select from. Compared to chemical mutagens and transposable elements, gamma rays offer a more controlled and predictable mutagenesis process, minimizing the risk of introducing unwanted mutations and disruptions. The accumulated knowledge and experience in gamma ray induced mutagenesis, along with its alignment with sustainable agriculture principles, make it a powerful tool for crop breeders seeking to enhance crop traits and address global food security challenges.

2.2 Examples of successful applications of gamma ray induced mutagenesis in crop improvement

The versatility and success of gamma ray induced mutagenesis are evident in its application across a wide range of crops. Various studies have demonstrated its effectiveness in crop improvement, leading to the development of new and improved varieties in rice, wheat, barley, soybean, and tomato [2].

In rice, gamma ray induced mutagenesis has been instrumental in enhancing important agronomic traits. For instance, researchers have used this technique to develop rice varieties with improved yield potential, disease resistance, and drought tolerance [6]. These newly created varieties hold great promise for addressing the challenges faced by rice farmers, such as increasing productivity and combating emerging diseases and climate-related stresses.

Similarly, gamma ray induced mutagenesis has been applied in wheat breeding to improve gluten quality and baking properties. By subjecting wheat plants to gamma rays, breeders have successfully generated new wheat varieties with enhanced gluten characteristics, leading to improved bread-making quality and consumer appeal [7]. This advancement not only benefits wheat producers but also contributes to meeting the demands of consumers for high-quality baked goods.

In the case of soybean, gamma ray induced mutagenesis has been utilized to develop varieties with improved nutritional profiles. Researchers have employed this technique to create soybean varieties with increased protein content, addressing the needs of industries that rely on soybean as a protein source [8]. Additionally, the reduction of anti-nutritional factors through gamma ray induced mutagenesis enhances the nutritional value of soybean, making it a more desirable and beneficial crop.

The successful application of gamma ray induced mutagenesis in these crops highlights its potential to address specific challenges and target desired traits. By inducing mutations in the DNA, this technique allows breeders to explore and exploit the genetic diversity within a crop species, leading to the development of novel varieties with improved characteristics.

It is important to note that gamma ray induced mutagenesis is not a standalone solution but rather a valuable tool in the broader context of crop improvement. It can be combined with other breeding strategies, such as traditional breeding methods and advanced molecular techniques, to further enhance the efficiency and precision of trait selection.

In addition to its practical applications, gamma ray induced mutagenesis contributes to fundamental research in plant genetics and genomics. The induced mutations serve as valuable resources for studying gene function and understanding the molecular mechanisms underlying important traits [2]. The insights gained from such studies can inform future breeding efforts and aid in the development of improved crop varieties.

In conclusion, gamma ray induced mutagenesis has proven to be a successful and versatile tool in crop improvement. Its application in various crops, including rice, wheat, barley, soybean, and tomato, has led to the development of novel varieties with improved traits such as yield, disease resistance, drought tolerance, gluten quality, and nutritional content. By harnessing the power of gamma rays to induce mutations, breeders can unlock the genetic potential within crop species and address the challenges faced by farmers and consumers. Furthermore, gamma ray induced mutagenesis facilitates research in plant genetics and genomics, contributing to the advancement of scientific knowledge in the field.

2.3 Potential for new applications of gamma ray induced mutagenesis in climate-smart agriculture

Gamma ray induced mutagenesis holds significant potential in the realm of climate-smart agriculture, offering a promising approach to create genetic variants with desirable traits such as drought tolerance, heat tolerance, and resistance to pests and diseases [2]. This technique has already demonstrated its effectiveness in developing new varieties of wheat and tomato with improved traits, showcasing its applicability in addressing the challenges of climate change and global food security.

In the face of increasing water scarcity and drought events, the development of crops with enhanced drought tolerance is crucial for sustainable agriculture. Gamma ray induced mutagenesis has been utilized to create new varieties of wheat with improved drought tolerance and yield stability under water-limited conditions [9]. These new wheat varieties are equipped with genetic modifications that enable them to better withstand water stress, ensuring more reliable yields and reducing the vulnerability of farmers to drought-related losses. By harnessing the power of gamma rays to induce mutations, breeders can access the genetic diversity within wheat populations and select for traits that confer greater resilience to water scarcity.

Another critical aspect of climate-smart agriculture is the development of crops with enhanced heat tolerance. Rising global temperatures pose a significant threat to crop productivity, making the creation of heat-tolerant varieties imperative. Gamma ray induced mutagenesis has been applied to tomato breeding, resulting in the creation of new varieties that exhibit improved heat tolerance and fruit quality [10]. These heat-tolerant tomatoes are capable of withstanding high temperature conditions, maintaining their productivity and fruit quality even under heat stress. By incorporating heat tolerance traits into crops through gamma ray induced mutagenesis, farmers can mitigate the negative impact of heat waves and ensure stable crop production.

Furthermore, gamma ray induced mutagenesis can contribute to the development of crops with enhanced resistance to pests and diseases. In the context of climate change, the prevalence and intensity of pests and diseases are expected to increase, posing significant challenges to agricultural productivity. By inducing mutations in targeted genes, gamma ray induced mutagenesis can generate genetic variations that confer resistance to specific pests and diseases, reducing the reliance on chemical pesticides and promoting sustainable pest management strategies. While specific examples of gamma ray induced mutagenesis for pest and disease resistance may not be cited here, the technique's potential in this area is well-established in the literature.

The successful application of gamma ray induced mutagenesis in enhancing drought tolerance, heat tolerance, and potentially pest and disease resistance highlights its potential to contribute to climate-smart agriculture. By utilizing this technique, breeders gain access to a vast pool of genetic diversity and can select and develop crop varieties that are better suited to the changing climate conditions. These genetically improved crops have the potential to enhance agricultural productivity, ensure food security, and promote sustainable farming practices.

It is important to note that gamma ray induced mutagenesis is not a standalone solution but rather a part of a comprehensive approach to climate-smart agriculture. It can be integrated with other strategies, such as precision breeding techniques and sustainable agricultural practices, to optimize the development and deployment of climate-resilient crop varieties.

Gamma ray induced mutagenesis thus has the potential to revolutionize climate-smart agriculture by enabling the creation of new genetic variants with traits such as drought tolerance, heat tolerance, and resistance to pests and diseases. The examples of wheat and tomato varieties developed through gamma ray induced mutagenesis illustrate the technique's effectiveness in breeding crops with improved traits. By harnessing the power of gamma rays, breeders can enhance the resilience and productivity of crops, contributing to global food security efforts in the face of climate change. Continued research and application of gamma ray induced mutagenesis, combined with other innovative strategies, hold promise for a sustainable and resilient agricultural future.

3. Advancements in gamma ray induced mutagenesis for crop improvement

3.1 High-throughput screening techniques for identifying desirable traits

The field of gamma ray induced mutagenesis has benefited from recent advancements in high-throughput screening techniques, which have revolutionized the identification and selection of desirable traits resulting from this method. These techniques enable the efficient screening of large populations of plants to identify specific traits such as disease resistance, yield potential, and quality attributes [2]. Among these techniques, phenotyping has emerged as a powerful tool in gamma ray induced mutagenesis research, utilizing imaging technologies to collect precise data on plant growth and development [11]. Additionally, other high-throughput screening techniques, including genomic selection, transcriptomics, and metabolomics, contribute to the comprehensive characterization of plant traits and facilitate the identification of valuable genetic variants [2].

Phenotyping, as a high-throughput screening technique, plays a crucial role in gamma ray induced mutagenesis research. It involves the non-destructive and automated collection of data on various plant characteristics, allowing for the rapid evaluation of large plant populations. Advanced imaging technologies, such as hyperspectral imaging and 3D imaging, capture detailed information on plant morphology, leaf structure, and canopy architecture [11]. By analyzing these imaging data, researchers can identify plants with desired traits, such as increased disease resistance or improved productivity, and select them for further breeding or genetic analysis. The implementation of phenotyping techniques in gamma ray induced mutagenesis accelerates the screening process, increases the efficiency of trait selection, and enhances the precision of crop improvement efforts.

In addition to phenotyping, other high-throughput screening techniques contribute to the comprehensive characterization of plant traits resulting from gamma ray induced mutagenesis. Genomic selection, for instance, involves the analysis of molecular markers across the entire genome to predict the performance of plants based on their genetic makeup [2]. By examining the genetic variations induced by gamma ray irradiation, researchers can identify specific markers associated with desirable traits and use them for targeted breeding programs. This approach enables the selection of plants with superior genetic potential, enhancing the efficiency of gamma ray induced mutagenesis in crop improvement.

Furthermore, transcriptomics and metabolomics provide valuable insights into the genetic and biochemical changes that occur in plants as a result of gamma ray induced mutagenesis. Transcriptomics involves the analysis of gene expression patterns, allowing researchers to identify genes that are upregulated or downregulated in response to mutagenesis [2]. By understanding the molecular mechanisms underlying desirable traits, breeders can develop a deeper understanding of the genetic basis of these traits and devise targeted strategies for trait improvement. Metabolomics, on the other hand, focuses on the analysis of metabolic profiles, providing information on the biochemical composition of plants [2]. This approach helps identify changes in metabolite levels that are associated with desirable traits, such as increased nutritional content or enhanced stress tolerance. The integration of transcriptomics and metabolomics data with gamma ray induced mutagenesis research allows for a comprehensive understanding of the complex interactions between genes, proteins, and metabolites, leading to more informed breeding decisions.

These high-throughput screening techniques have transformed the field of gamma ray induced mutagenesis, enabling researchers to efficiently and effectively identify desirable traits and select plants with the highest potential for crop improvement. By combining phenotyping, genomic selection, transcriptomics, and metabolomics, breeders can gain a holistic understanding of the genetic variations induced by gamma ray irradiation and their impact on plant traits. This knowledge enhances the precision and efficiency of crop improvement programs, leading to the development of new and improved crop varieties that address the challenges of modern agriculture, such as increasing productivity, enhancing nutritional quality, and improving stress tolerance.

In conclusion, recent advancements in high-throughput screening techniques have revolutionized the field of gamma ray induced mutagenesis. Phenotyping, genomic selection, transcriptomics, and metabolomics provide valuable tools for the identification and selection of desirable traits resulting from gamma ray irradiation. These techniques enable the rapid screening of large plant populations, the analysis of genetic and molecular changes, and the comprehensive characterization of plant traits. By integrating these high-throughput screening techniques into gamma ray induced mutagenesis research, breeders can enhance the precision and efficiency of crop improvement efforts, leading to the development of new and improved crop varieties that address the challenges of modern agriculture.

3.2 Use of molecular markers for tracking mutations

In the field of gamma ray induced mutagenesis, the utilization of molecular markers has gained significant importance in tracking mutations and identifying desirable traits. Molecular markers, including single nucleotide polymorphisms (SNPs) and microsatellites, play a crucial role in identifying genetic variations within plant populations [2]. By leveraging these markers, researchers can track mutations induced by gamma ray irradiation and identify specific genetic variations associated with desirable traits. This information is invaluable for selecting plants with these traits for further breeding programs, facilitating the development of improved crop varieties [4].

Single nucleotide polymorphisms (SNPs) are one of the most commonly used molecular markers in gamma ray induced mutagenesis research. SNPs are variations in a single nucleotide within the DNA sequence, and their presence can be detected through genotyping techniques. By analyzing the SNPs present in a plant population, researchers can identify specific genetic variations associated with desirable traits resulting from gamma ray induced mutagenesis. This information allows for the selection of plants that carry these beneficial genetic variations, enabling breeders to focus their efforts on individuals with the highest potential for crop improvement.

Microsatellites, also known as simple sequence repeats (SSRs), are another type of molecular marker widely used in gamma ray induced mutagenesis research.

Microsatellites are short, repetitive DNA sequences that exhibit variations in the number of repeats between individuals. These variations can be detected through polymerase chain reaction (PCR) amplification and subsequent fragment analysis. By examining the microsatellite variations within a plant population, researchers can identify the presence of genetic mutations induced by gamma ray irradiation. Moreover, microsatellite markers facilitate the identification of specific genetic variations associated with desirable traits, aiding in the selection of plants with these traits for further breeding programs.

The use of molecular markers in gamma ray induced mutagenesis research has several advantages. Firstly, they provide a means to track and identify genetic variations resulting from mutagenesis. This allows researchers to identify the specific mutations responsible for desirable traits, facilitating the selection of plants with those traits for subsequent breeding efforts. Molecular markers also enable the assessment of genetic diversity within plant populations, providing valuable information for crop improvement programs. By analyzing the genetic diversity, breeders can identify plants with unique combinations of genetic variations, increasing the chances of discovering novel and valuable traits.

Furthermore, molecular markers offer a more efficient and targeted approach to crop improvement in comparison to traditional breeding methods. Their use allows for the selection of plants with the desired traits at an early stage, reducing the time and resources required for breeding programs. By focusing on individuals with specific genetic variations, breeders can streamline their efforts and accelerate the development of improved crop varieties.

It is worth noting that the use of molecular markers in gamma ray induced mutagenesis is not without challenges. The identification and validation of marker-trait associations require extensive genetic and phenotypic data, as well as rigorous statistical analyses. Additionally, the availability of marker information for specific traits can be limited, particularly for less-studied crops. However, with advancements in genomics and bioinformatics, the development of high-density marker platforms and databases has facilitated the identification and utilization of molecular markers for crop improvement.

The use of molecular markers, such as SNPs and microsatellites, has become increasingly important in gamma ray induced mutagenesis research. These markers enable the tracking of mutations and the identification of genetic variations associated with desirable traits resulting from gamma ray irradiation. By leveraging molecular markers, researchers can select plants with specific genetic variations for further breeding programs, expediting the development of improved crop varieties. Despite challenges associated with marker identification and validation, the use of molecular markers offers a targeted and efficient approach to crop improvement, enhancing the precision and effectiveness of gamma ray induced mutagenesis in agriculture.

3.3 New tools and technologies for analyzing and characterizing mutations

In the realm of gamma ray induced mutagenesis, the emergence of new tools and technologies has revolutionized the analysis and characterization of mutations. Two such advancements, namely next-generation sequencing (NGS) and CRISPR/Cas9 gene editing, have significantly contributed to understanding the effects of gamma ray induced mutagenesis and creating new genetic variants with desirable traits.

Next-generation sequencing (NGS) has transformed the field of genomics by enabling the rapid and cost-effective sequencing of large amounts of DNA. This technology has proven invaluable in the analysis of mutations resulting from gamma ray induced mutagenesis. By sequencing the genomes of mutated plants, researchers can identify and characterize the mutations at a comprehensive scale. NGS allows for the detection of different types of mutations, such as single nucleotide polymorphisms (SNPs), insertions, deletions, and structural rearrangements, providing a comprehensive view of the genetic changes induced by gamma ray irradiation [2]. Furthermore, NGS facilitates the identification of mutations in both coding and non-coding regions of the genome, thereby enabling researchers to explore their potential effects on gene expression and function. This information is crucial for understanding the underlying mechanisms of desirable traits resulting from gamma ray induced mutagenesis and for guiding subsequent breeding efforts.

CRISPR/Cas9 gene editing is another groundbreaking technology that has revolutionized genetic research and crop improvement. It allows for precise and targeted modifications of specific genes in plants, enabling the creation of new genetic variants with desired traits. The CRISPR/Cas9 system consists of a guide RNA that directs the Cas9 enzyme to a specific DNA sequence, where it introduces targeted modifications, such as insertions, deletions, or substitutions. In the context of gamma ray induced mutagenesis, CRISPR/Cas9 can be used to introduce specific mutations into genes of interest, mimicking the mutations induced by gamma ray irradiation [4]. This targeted approach offers several advantages over traditional mutagenesis methods, as it allows for precise modifications without introducing a large number of random mutations. By using CRISPR/Cas9 gene editing, researchers can create new genetic variants with desirable traits and investigate the functional consequences of specific mutations induced by gamma ray irradiation.

The combination of NGS and CRISPR/Cas9 gene editing has the potential to revolutionize gamma ray induced mutagenesis research and crop improvement. NGS provides a powerful tool for genome-wide analysis of mutations, allowing for the identification of specific genetic variations associated with desirable traits. By integrating NGS data with phenotypic information, researchers can gain insights into the functional consequences of mutations and their impact on plant traits. This knowledge is essential for identifying key genes and regulatory elements involved in desirable traits and for developing targeted strategies to enhance the expression of these traits in crops.

Additionally, CRISPR/Cas9 gene editing complements NGS by enabling the precise manipulation of specific genes in plants. By introducing or modifying mutations in target genes, researchers can validate the functional significance of specific mutations identified through NGS analysis. CRISPR/Cas9 gene editing also offers the possibility of directly creating new genetic variants with desirable traits [12] without relying solely on random mutagenesis induced by gamma ray irradiation. This targeted approach accelerates the breeding process by allowing for the creation of specific genetic variations that are known to contribute to desirable traits.

In conclusion, the advent of new tools and technologies such as NGS and CRISPR/Cas9 gene editing has transformed gamma ray induced mutagenesis research and crop improvement. NGS enables comprehensive analysis of mutations induced by gamma ray irradiation, providing insights into their effects on gene expression and function. CRISPR/Cas9 gene editing allows for targeted modifications of specific genes, facilitating the creation of new genetic variants with desirable traits. By integrating NGS analysis with CRISPR/Cas9 gene editing, researchers can gain a deeper understanding of the functional consequences of specific mutations and develop precise strategies for crop improvement. These advancements hold great promise for enhancing the efficiency and precision of gamma ray induced mutagenesis and driving the development of improved crop varieties to meet the challenges of modern agriculture.

3.4 Integration with climate-smart agriculture practices

The integration of gamma ray induced mutagenesis with climate-smart agriculture practices presents a promising approach for developing new genetic variants with desirable traits that can withstand the challenges imposed by changing climatic conditions. By combining high-throughput screening techniques, molecular markers, and gene editing technologies, researchers can identify and modify specific genes associated with traits such as drought tolerance, heat tolerance, and resistance to pests and diseases [2]. This integration holds great potential for addressing global food security challenges and enhancing agricultural sustainability.

Climate-smart agriculture aims to increase agricultural productivity, enhance resilience to climate change, and reduce greenhouse gas emissions. It encompasses a range of practices and technologies that promote sustainable and efficient use of resources while adapting to and mitigating the effects of climate change. Gamma ray induced mutagenesis can be leveraged within the framework of climate-smart agriculture to develop crop varieties that are better adapted to the specific challenges imposed by a changing climate.

High-throughput screening techniques play a crucial role in the integration of gamma ray induced mutagenesis with climate-smart agriculture. These techniques allow for the rapid screening of large populations of mutated plants, enabling the identification of individuals with desirable traits. By subjecting these populations to various stress conditions, such as drought, heat, or pest infestation, researchers can identify plants that exhibit enhanced tolerance or resistance. This screening process helps to identify genetic variations induced by gamma ray irradiation that are associated with the desired traits, providing valuable information for subsequent breeding and selection programs.

Molecular markers, such as single nucleotide polymorphisms (SNPs) and microsatellites, are essential tools in the integration of gamma ray induced mutagenesis with climate-smart agriculture. These markers enable the identification and tracking of specific genetic variations associated with desirable traits. By analyzing the genetic diversity within plant populations using molecular markers, researchers can identify marker-trait associations and select plants with the desired traits for further breeding programs. This targeted approach increases the efficiency and precision of crop improvement efforts, allowing for the development of new genetic variants that are better adapted to specific climatic challenges.

Furthermore, gene editing technologies, particularly CRISPR/Cas9, offer immense potential in the integration of gamma ray induced mutagenesis with climate-smart agriculture. With gene editing, specific genes can be modified or precisely targeted to introduce desirable traits or enhance existing ones. For instance, researchers can identify key genes associated with drought tolerance and use CRISPR/Cas9 gene editing to introduce specific mutations that enhance the plant's ability to withstand water scarcity. Similarly, genes involved in heat tolerance or resistance to pests and diseases can be targeted and modified to improve crop resilience under challenging climatic conditions. Gene editing technologies offer a powerful tool for precisely engineering desired traits, providing an additional avenue for crop improvement within the context of climate-smart agriculture.

By integrating gamma ray induced mutagenesis with climate-smart agriculture practices, it becomes possible to develop new crop varieties that are better adapted to changing climatic conditions. These improved varieties can contribute to addressing global food security challenges by increasing agricultural productivity and resilience. Moreover, they can reduce the reliance on external inputs, such as water, pesticides, and fertilizers, thereby promoting sustainable and resource-efficient agricultural systems.

In conclusion, the integration of gamma ray induced mutagenesis with climate-smart agriculture practices offers a promising approach for crop improvement in the face of changing climatic conditions. The combination of high-throughput screening techniques, molecular markers, and gene editing technologies allows for the identification and modification of specific genes associated with desirable traits. By developing crop varieties that exhibit enhanced tolerance to drought, heat, and pests, it is possible to enhance agricultural productivity, resilience, and sustainability. This integrated approach holds significant potential for addressing global food security challenges and ensuring the long-term viability of agricultural systems in a changing climate.

4. Challenges associated with gamma ray induced mutagenesis for crop improvement

4.1 Balancing trait improvements with potential negative effects on crop yield or quality

Although gamma ray induced mutagenesis holds great promise for crop improvement, it is crucial to address the challenge of balancing trait improvements with potential negative effects on crop yield or quality. While this mutagenesis technique can generate novel genetic variants with desirable traits, it can also introduce unintended negative effects, such as reduced yield or compromised quality [4]. Therefore, a thorough evaluation of the effects of mutations on crop performance is essential, and only mutations that positively impact crop yield and quality should be selected for further development.

When conducting gamma ray induced mutagenesis experiments, it is important to consider that not all mutations will result in beneficial traits. In fact, the majority of mutations induced by gamma ray irradiation are likely to be neutral or have deleterious effects. These mutations can disrupt essential genes, regulatory regions, or metabolic pathways, leading to negative impacts on crop growth, development, and productivity. Therefore, it is crucial to carefully evaluate the phenotypic consequences of mutations and select only those that confer a clear advantage in terms of yield and quality.

To evaluate the effects of mutations on crop performance, rigorous phenotypic screening should be conducted. This involves assessing various agronomic traits, such as plant height, flowering time, seed set, fruit size, and nutritional composition, among others. By comparing mutated plants with the original cultivar or control plants, researchers can identify mutations that positively impact these traits. It is important to conduct multiple replicated trials to account for environmental variations and ensure the consistency of the observed phenotypic effects.

In addition to phenotypic screening, molecular techniques can provide valuable insights into the underlying genetic changes induced by gamma ray mutagenesis. High-throughput sequencing technologies, such as next-generation sequencing (NGS), can be employed to identify and characterize the specific mutations present in the genomes of mutated plants. This genomic information can help researchers understand the functional consequences of mutations and their potential impacts on crop yield and quality. By integrating phenotypic and genomic data, researchers can identify associations between specific mutations and desirable traits, facilitating the selection of promising genetic variants for further breeding programs.

Furthermore, it is crucial to consider the potential negative effects of mutations on crop quality. Gamma ray induced mutagenesis may inadvertently impact the nutritional composition, flavor, texture, or other quality attributes of crops. For instance, a mutation that enhances drought tolerance may result in a decrease in crop quality, such as reduced sugar content or altered taste. Therefore, comprehensive quality assessments should be conducted to ensure that the selected mutations do not compromise crop quality traits that are important for consumer acceptance and marketability.

To mitigate the risk of negative effects on crop yield or quality, it is essential to utilize a combination of strategies. One approach is to prioritize mutations that have been previously identified and validated in related crops or species. This knowledge can help guide the selection of mutations that are more likely to result in desirable traits without detrimental effects on yield or quality. Additionally, incorporating gene editing technologies, such as CRISPR/Cas9, can provide a targeted and precise approach to introduce specific mutations associated with desirable traits, minimizing the potential for unintended negative effects.

In conclusion, while gamma ray induced mutagenesis offers opportunities for creating novel genetic variants with desirable traits, it is crucial to carefully evaluate the effects of mutations on crop performance, yield, and quality. Rigorous phenotypic screening, combined with genomic analyses, can help identify mutations that positively impact crop traits. Additionally, assessing potential negative effects on crop quality is essential to ensure consumer acceptance and marketability. By employing a comprehensive evaluation strategy and integrating gene editing technologies, researchers can strike a balance between trait improvements and potential negative effects, ultimately selecting mutations that contribute positively to crop yield and quality in the context of gamma ray induced mutagenesis.

4.2 Ethical considerations of using mutagenic agents in agriculture

One of the challenges associated with gamma ray induced mutagenesis in agriculture pertains to the ethical considerations surrounding the use of mutagenic agents. There are concerns among some individuals regarding the safety and potential long-term effects of employing mutagenic agents on crops, as well as the possible impacts on human health and the environment [3]. Therefore, it is crucial to conduct a thorough evaluation of the safety aspects associated with the use of mutagenic agents in agriculture and implement appropriate safety measures to safeguard human health and the environment.

To address the ethical concerns surrounding gamma ray induced mutagenesis, it is essential to prioritize the safety evaluation of mutagenic agents. Rigorous testing protocols should be established to assess the potential risks associated with the use of gamma rays as mutagenic agents. These protocols should include comprehensive studies on the effects of gamma ray irradiation on the genetic stability of crops, as well as investigations into potential impacts on non-target organisms and the overall ecosystem.

Human health considerations are of utmost importance in evaluating the safety of mutagenic agents. It is essential to determine whether the consumption of crops derived from gamma ray induced mutagenesis poses any risks to human health. Toxicological studies should be carried out to examine the potential accumulation of harmful substances and the effects on nutritional composition or allergenicity of mutated crops. Additionally, studies should investigate the potential presence of unintended toxins or anti-nutritional compounds resulting from the mutagenesis process.

Environmental safety is another crucial aspect that needs to be addressed. The possible effects of gamma ray induced mutagenesis on non-target organisms, including beneficial insects, pollinators, and soil microorganisms, should be thoroughly investigated. Assessments should be conducted to determine the potential ecological impacts of released mutant crops on biodiversity and ecosystem functioning. Furthermore, studies should examine the potential for gene flow from mutated crops to wild relatives, considering the potential consequences for biodiversity and the development of herbicide-resistant weeds.

To mitigate potential risks and ensure the responsible use of mutagenic agents, appropriate safety measures should be implemented. Regulatory frameworks and guidelines should be established to govern the use of gamma ray induced mutagenesis in agriculture. These frameworks should include stringent safety assessments and risk management protocols to minimize any potential adverse effects. Additionally, comprehensive labeling and traceability systems should be put in place to ensure transparency and enable consumer choice.

Engagement with stakeholders, including scientists, policymakers, farmers, and consumers, is crucial in addressing the ethical considerations associated with gamma ray induced mutagenesis. Public awareness campaigns and educational initiatives can help foster an understanding of the benefits, risks, and safety measures associated with mutagenic agents in agriculture. Open dialogue and transparency regarding the research, development, and implementation of mutagenesis techniques can help build trust and ensure that ethical concerns are adequately addressed.

In conclusion, the ethical considerations surrounding the use of mutagenic agents, such as gamma rays, in agriculture require careful evaluation and appropriate safety measures. Addressing concerns related to human health and environmental impacts is paramount. Rigorous safety evaluations, including toxicological studies, assessments of genetic stability, and investigations into potential ecological effects, should be conducted. Regulatory frameworks, labeling systems, and traceability measures should be established to ensure responsible use and transparency. Engaging stakeholders through education and dialogue is essential for building trust and addressing ethical concerns. By prioritizing safety and implementing appropriate measures, the ethical challenges associated with gamma ray induced mutagenesis can be effectively managed in the context of agricultural practices.

4.3 Safety measures for protecting the environment and human health

In order to alleviate concerns surrounding the safety of mutagenic agents in agriculture, it is imperative to implement comprehensive safety measures that prioritize the protection of the environment and human health. These measures encompass a range of practices, including the utilization of appropriate protective equipment and adherence to specific procedures for safe handling of mutagenic agents. Additionally, the establishment of regulations and guidelines governing the use of mutagenic agents in agriculture can further ensure safety [3]. Furthermore, it is crucial to conduct diligent monitoring of the effects of mutagenic agents on the environment and promptly implement measures to mitigate any potential negative impacts.

The implementation of safety measures in the use of mutagenic agents is essential to safeguard both the environment and human health. Protective equipment, such as gloves, masks, and lab coats, should be used when handling mutagenic agents to prevent direct exposure and minimize the risk of contamination. Furthermore, designated areas and facilities should be established for the storage, handling, and disposal of mutagenic agents to prevent accidental release into the environment. Adequate training and education should be provided to individuals working with mutagenic agents to ensure proper handling techniques and awareness of potential risks.

Regulations and guidelines play a crucial role in promoting the safe use of mutagenic agents in agriculture. Government bodies and regulatory agencies should establish comprehensive frameworks that outline the requirements and safety standards for the use of mutagenic agents. These frameworks should cover aspects such as permissible dosage levels, recommended application methods, and appropriate containment measures. Regular inspections and audits can help ensure compliance with these regulations, and penalties should be enforced for non-compliance to deter unsafe practices.

Monitoring the effects of mutagenic agents on the environment is essential for identifying and addressing any potential negative impacts. Environmental monitoring programs can be implemented to assess the presence and persistence of mutagenic agents in soil, water, and air. These programs can also help identify any changes in the composition and diversity of non-target organisms in the surrounding ecosystem. By regularly monitoring and analyzing the data, any adverse effects can be detected early on, allowing for timely intervention and the implementation of appropriate mitigation measures.

Mitigation measures should be established to minimize any negative impacts associated with the use of mutagenic agents. These measures may include the development of containment strategies to prevent the spread of mutated genetic material beyond the intended target crop. For example, physical barriers, isolation distances, or time-based planting schedules can be employed to reduce the potential for gene flow and the establishment of unintended populations in the environment. Additionally, strategies for the safe disposal of mutagenic agents and their by-products should be implemented to prevent contamination and minimize any potential long-term effects.

Collaboration and information sharing among researchers, regulatory bodies, and stakeholders are crucial for the effective implementation of safety measures. Open communication channels facilitate the exchange of knowledge, best practices, and lessons learned regarding the safe use of mutagenic agents. Collaborative efforts can lead to the development of standardized protocols, risk assessment methodologies, and safety guidelines that encompass the concerns of all stakeholders involved.

In conclusion, addressing concerns about the safety of mutagenic agents in agriculture requires the implementation of robust safety measures. These measures encompass the use of appropriate protective equipment, adherence to handling procedures, and the establishment of regulations and guidelines. Monitoring the effects of mutagenic agents on the environment is essential, as is the implementation of mitigation measures to minimize negative impacts. Collaboration and information sharing among stakeholders are key to ensuring the safe and responsible use of mutagenic agents in agriculture. By implementing comprehensive safety measures and fostering a culture of responsible practice, the concerns regarding the safety of mutagenic agents can be effectively addressed, safeguarding both the environment and human health.

4.4 Considerations for climate-smart agriculture

In the context of climate-smart agriculture, the utilization of gamma ray induced mutagenesis requires a careful assessment of the potential impacts that new genetic variants may have on both the environment and crop performance. While the development of new genetic variants aimed at enhancing drought tolerance holds promise, it is crucial to consider the unintended negative effects that these variants may have on crop yield and quality under normal conditions [2]. Therefore, it is necessary to conduct thorough evaluations of the effects of new genetic variants on crop performance across a range of environmental conditions and to select only those variants that demonstrate a positive impact on crop yield and quality.

When developing new genetic variants for climate-smart agriculture, it is important to consider the potential consequences they may have on crop performance beyond their intended purpose. While the focus may be on enhancing drought tolerance, it is essential to assess the overall impact of these variants on crop yield and quality under normal growing conditions. The introduction of new genetic variants can inadvertently lead to negative effects, such as reduced yield potential, altered growth patterns, or compromised nutritional quality. Therefore, comprehensive evaluations are necessary to ensure that the benefits of enhanced drought tolerance are not outweighed by unintended negative consequences.

Evaluating the effects of new genetic variants on crop performance requires rigorous testing under various environmental conditions. Field trials should be conducted to assess the performance of mutant crops alongside traditional varieties across different agro-climatic regions. Multiple parameters, such as yield potential, growth characteristics, disease resistance, and nutritional composition, should be carefully monitored and compared. This evaluation process helps identify genetic variants that not only exhibit improved drought tolerance but also maintain or enhance overall crop performance under normal conditions.

The selection of genetic variants with positive impacts on crop yield and quality is crucial for the success of climate-smart agriculture. Through meticulous evaluation, only those variants that consistently exhibit improved performance across multiple trials and environmental conditions should be selected for further development and deployment. This selection process ensures that the chosen genetic variants provide sustainable benefits without compromising crop productivity or quality. Additionally, the inclusion of diverse germplasm in the evaluation process helps identify a wider range of genetic variants with desirable traits, increasing the chances of success in breeding programs.

To enhance the selection process, the integration of advanced technologies, such as genomic and phenomic analyses, can provide valuable insights into the performance and characteristics of new genetic variants. These technologies allow for a more comprehensive understanding of the genetic makeup, gene expression patterns, and physiological responses of mutant crops. By combining traditional breeding methods with cutting-edge technologies, breeders can make informed decisions regarding the selection of genetic variants that offer the desired traits without sacrificing overall crop performance.

Collaboration and knowledge exchange among researchers, breeders, and farmers are essential for the successful implementation of climate-smart agriculture using gamma ray induced mutagenesis. By sharing information and experiences, stakeholders can collectively identify genetic variants that possess the desired traits and evaluate their performance across diverse environments and cropping systems. This collaborative approach helps ensure that the selected genetic variants are applicable and effective in real-world agricultural settings.

In conclusion, the development and deployment of new genetic variants using gamma ray induced mutagenesis for climate-smart agriculture necessitate careful consideration of their potential impacts on the environment and crop performance. It is essential to evaluate the effects of these variants on crop yield and quality under various environmental conditions and select only those that demonstrate positive impacts. Rigorous testing, including field trials and the incorporation of advanced technologies, aids in the identification of genetic variants that offer enhanced drought tolerance without compromising overall crop performance. Collaborative efforts among stakeholders further enhance the selection process and facilitate the successful implementation of climate-smart agriculture. By ensuring that new genetic variants have a positive impact on crop productivity and quality, gamma ray induced mutagenesis can contribute to sustainable and resilient agricultural systems.

5. Future directions of gamma ray induced mutagenesis for crop improvement

5.1 Potential for further advancements in the use of gamma rays to induce mutations in crops

The use of gamma rays to induce mutations in crops for crop improvement holds significant potential for further advancements. Ongoing developments in radiation technology present opportunities to enhance the precision and effectiveness of this technique. These advancements may enable scientists to target specific regions of the DNA more precisely, thereby facilitating the creation of more targeted mutations and the development of new genetic variants with even more desirable traits [2].

Advancements in radiation technology can revolutionize the process of gamma ray induced mutagenesis. Traditional approaches involve exposing seeds or plant tissues to gamma radiation, which leads to random mutations throughout the genome. However, with the emergence of more precise radiation technologies, such as ion beams or CRISPR-based techniques, it becomes possible to direct the mutagenic effects to specific regions of the DNA. This targeted approach allows for greater control over the genetic modifications, facilitating the creation of desired traits while minimizing unintended changes in other genomic regions. By harnessing these advancements, scientists can optimize the efficiency and effectiveness of gamma ray induced mutagenesis for crop improvement.

In addition to advancements in radiation technology, progress in high-throughput screening and molecular marker technologies contributes to the further development of gamma ray induced mutagenesis. High-throughput screening techniques enable the rapid and simultaneous screening of a large number of mutant plants for specific traits of interest. This accelerates the identification of desirable traits resulting from gamma ray induced mutagenesis, allowing breeders to more efficiently select and advance promising genetic variants. By integrating high-throughput screening with molecular marker technologies, such as next-generation sequencing or DNA microarrays, researchers gain valuable insights into the genetic changes induced by gamma rays. These technologies enable the identification and characterization of specific mutations associated with desirable traits, providing a more comprehensive understanding of the genetic basis underlying trait improvement [4].

The integration of high-throughput screening and molecular marker technologies with gamma ray induced mutagenesis offers several advantages. Firstly, it expedites the identification of desired traits, allowing breeders to save time and resources by focusing on plants with the most promising characteristics. Secondly, it facilitates the selection of genetic variants with multiple desirable traits, as the simultaneous screening of a large population enables the identification of plants exhibiting a combination of beneficial traits. Thirdly, it provides valuable insights into the genetic mechanisms underlying trait improvement, enabling a more targeted and informed approach to crop breeding. By capitalizing on these advancements, scientists can enhance the efficiency and precision of gamma ray induced mutagenesis, leading to the development of improved crop varieties.

Despite these advancements, it is important to note that gamma ray induced mutagenesis should be employed within a comprehensive breeding program. It is not a standalone solution but rather a tool that complements other breeding methods, such as conventional breeding or genetic engineering. Combining gamma ray induced mutagenesis with other approaches allows for the integration of diverse genetic resources and the exploitation of various mechanisms for trait improvement. This integrated approach maximizes the potential for developing crop varieties with enhanced traits, including increased yield, improved nutritional content, enhanced disease resistance, and superior adaptation to changing environmental conditions.

In conclusion, the use of gamma rays to induce mutations in crops for crop improvement holds significant potential for further advancements. Ongoing developments in radiation technology enable more precise targeting of specific regions of the DNA, leading to the creation of more targeted mutations and the development of new genetic variants with even more desirable traits. Advancements in high-throughput screening and molecular marker technologies further enhance the identification and characterization of desirable traits resulting from gamma ray induced mutagenesis. By integrating these advancements, scientists can optimize the efficiency and effectiveness of gamma ray induced mutagenesis, contributing to the development of improved crop varieties. It is crucial to employ gamma ray induced mutagenesis within a comprehensive breeding program to fully harness its potential and achieve sustainable and resilient agriculture.

5.2 Integration of gamma ray induced mutagenesis with other techniques for crop improvement

The integration of gamma ray-induced mutagenesis with other techniques for crop improvement holds tremendous potential for advancing plant breeding. By combining gamma ray-induced mutagenesis with techniques like gene editing and marker-assisted selection, researchers can explore new avenues for creating improved crop varieties. This integration may lead to even more precise and targeted mutations, resulting in the development of genetic variants with highly desirable traits [2].

One of the techniques that can be integrated with gamma ray-induced mutagenesis is CRISPR/Cas9 gene editing. CRISPR/Cas9 enables scientists to make specific modifications to the DNA sequence, allowing for precise genetic alterations. When combined with gamma ray-induced mutagenesis, this technique offers the potential to create more targeted mutations. By using CRISPR/Cas9 to introduce specific genetic changes in conjunction with gamma ray-induced mutagenesis, researchers can enhance the efficiency and precision of trait improvement. This integration may result in the development of new genetic variants with highly desirable traits, such as increased yield, enhanced nutritional content, or improved stress tolerance [2].

Furthermore, the integration of molecular markers with gamma ray-induced mutagenesis can enhance the identification and selection of desirable traits resulting from mutations. Molecular markers are specific DNA sequences that are associated with particular traits of interest. By using molecular marker technologies, researchers can efficiently screen a large number of plants and identify those with desired traits. This accelerates the breeding process by enabling the selection of plants with desirable characteristics at an early stage. Molecular markers can also facilitate the tracking of genetic variations induced by gamma ray-induced mutagenesis, allowing researchers to understand the genetic basis of the observed traits. By integrating molecular markers with gamma ray-induced mutagenesis, breeders can more effectively select and advance promising genetic variants, leading to the development of improved crop varieties [4].

The integration of gamma ray-induced mutagenesis with gene editing and molecular marker technologies offers several advantages. Firstly, it provides a more precise and targeted approach to trait improvement. By combining gamma ray-induced mutagenesis with gene editing, researchers can introduce specific genetic modifications, enhancing the efficiency of trait development. Secondly, the integration with molecular markers allows for efficient screening and selection of desired traits, enabling the identification of promising genetic variants at an early stage. This reduces the time and resources required for breeding programs. Thirdly, the integration of these techniques enables a more comprehensive understanding of the genetic mechanisms underlying trait improvement, leading to improved breeding strategies and outcomes.

It is important to note that the integration of gamma ray-induced mutagenesis with gene editing and molecular marker technologies should be conducted within appropriate regulatory frameworks and ethical considerations. The responsible use of these techniques ensures the safety and sustainability of crop improvement efforts.

In conclusion, the integration of gamma ray-induced mutagenesis with other techniques such as gene editing and molecular marker technologies opens up new possibilities for crop improvement. The integration of CRISPR/Cas9 gene editing with gamma ray-induced mutagenesis allows for the creation of more precise and targeted mutations, leading to the development of genetic variants with highly desirable traits. Similarly, the integration of molecular markers with gamma ray-induced mutagenesis enhances the identification and selection of desirable traits. These advancements contribute to more efficient and effective breeding programs, ultimately leading to the development of improved crop varieties. By responsibly integrating these techniques, researchers can continue to advance the field of plant breeding and contribute to sustainable and resilient agriculture [2, 4].

5.3 Assessing potential hazards in gamma ray-induced mutagenesis for climate-smart crop improvement: methodologies, criteria, and risk assessment

This review provides an overview of the application of gamma ray-induced mutagenesis in crop improvement within the context of climate-smart agriculture. While it highlights the potential benefits of this technique, such as the creation of new genetic variants with desirable traits, it is essential to have specific methodologies and criteria for assessing potential hazards associated with mutagenesis. While implementing this technique, a comprehensive risk assessment should be conducted, taking into account not only the direct health and environmental impacts but also the unintended consequences that may arise. These unintended consequences could include the development of secondary pests or the loss of genetic diversity in mutated crops.

To assess potential hazards, specific methodologies can be employed. These may include conducting field trials to evaluate the performance of mutated crops under different environmental conditions and monitoring their interactions with the ecosystem. Additionally, genetic analysis can be employed to assess changes in the genetic makeup of mutated crops and identify any potential risks associated with the introduction of new traits. Criteria for assessing potential hazards should be established, considering factors such as the severity and likelihood of the identified risks. This can involve evaluating the potential impacts on human health, ecosystem stability, and agronomic traits. Furthermore, the long-term effects of mutated crops should be monitored to identify any unforeseen consequences that may arise over time. Collaboration among stakeholders is crucial in ensuring the responsible and safe use of mutagenic agents in agriculture. Scientists, breeders, regulators, and farmers should work together to establish standardized protocols for risk assessment and monitoring. Ethical considerations regarding the safety of both humans and the environment should be integrated into the assessment process. Through collaboration and careful evaluation, the safe and responsible application of mutagenic agents in agriculture can be ensured.

5.4 Future challenges and opportunities for gamma ray induced mutagenesis in climate-smart agriculture

Gamma ray-induced mutagenesis holds great promise for developing crop varieties that are well-suited for climate-smart agriculture. However, it is essential to address the challenges associated with this technique to ensure its successful application. One of the primary concerns is the potential negative impact of mutations on crop yield or quality. While gamma ray-induced mutagenesis can generate beneficial genetic variations, it can also introduce undesirable changes that may adversely affect important agronomic traits [2].

To overcome these challenges, future research efforts should focus on developing new genetic variants that are not only adapted to changing climatic conditions but also exhibit high yield potential and good quality. This requires a comprehensive understanding of the genetic basis of desirable traits and the identification of specific mutations that confer these characteristics. By leveraging advancements in molecular biology and genomics, researchers can elucidate the underlying mechanisms of trait improvement and develop strategies to enhance the selection and breeding of improved crop varieties resulting from gamma ray-induced mutagenesis.

Another critical aspect that needs attention is the safety of using mutagenic agents in agriculture. Mutagenesis involves the deliberate induction of mutations in the genome, and it is essential to establish safety measures to protect human health and the environment. Strict regulations and guidelines should be in place to ensure the responsible and controlled use of gamma rays or other mutagenic agents. Risk assessment protocols should be developed and implemented to evaluate the potential hazards associated with the use of mutagenesis techniques and to mitigate any adverse effects. This includes monitoring and assessing the potential risks of unintended changes in the genome and the potential presence of any harmful substances in the mutated crops.

Furthermore, it is crucial to consider the potential long-term effects of using gamma ray-induced mutagenesis in agriculture. Continuous monitoring and evaluation of the performance and stability of the mutated crops are necessary to ensure their sustained benefits and assess any unintended consequences that may arise over time. Long-term studies can provide valuable insights into the stability of the desirable traits, potential gene flow to wild relatives, and the overall environmental impact of using mutagenesis techniques.

Collaboration between scientists, breeders, regulators, and other stakeholders is vital in addressing these challenges and ensuring the safe and effective use of gamma ray-induced mutagenesis in crop improvement. Transparent communication and knowledge sharing are essential for disseminating information about the benefits, risks, and regulatory frameworks associated with mutagenesis techniques. This collaboration can facilitate the development of robust safety measures, rigorous risk assessments, and effective monitoring systems.

In conclusion, gamma ray-induced mutagenesis presents significant opportunities for developing new crop varieties with desirable traits for climate-smart agriculture. However, challenges need to be addressed to realize the full potential of this technique. Future research should focus on developing genetic variants that are well-adapted to changing climatic conditions while maintaining high yield potential and good quality. Safety measures must be established and implemented to safeguard human health and the environment. By addressing these challenges and fostering collaboration among stakeholders, gamma ray-induced mutagenesis can contribute to the development of resilient crop varieties that meet the demands of a changing climate while ensuring the safety and sustainability of agricultural practices [2].

6. Conclusion

Gamma-ray-induced mutagenesis has long been recognized as a valuable tool for crop improvement, enabling the creation of new genetic variants with desirable traits. The integration of high-throughput screening, molecular marker technologies, and gene editing has expanded the possibilities of gamma ray-induced mutagenesis in crop improvement. These advancements allow for more efficient identification and selection of desirable traits resulting from mutations.

However, it is important to acknowledge the challenges associated with the use of mutagenic agents in agriculture. One concern is the potential negative impact on crop yield or quality. While gamma ray-induced mutagenesis can introduce beneficial genetic variations, it may also lead to undesirable changes that could adversely affect important agronomic traits.

Additionally, ethical considerations regarding the safety of using mutagenic agents in agriculture must be addressed. It is crucial to establish appropriate regulatory frameworks and safety measures to protect human health and the environment. Careful risk assessments should be conducted to evaluate any potential hazards associated with mutagenesis techniques, including the unintended changes in the genome and the presence of harmful substances in mutated crops.

In summary, gamma ray-induced mutagenesis, with its integration of high-throughput screening, molecular markers, and gene editing, offers great potential for crop improvement. However, challenges related to potential negative effects on crop yield or quality, as well as ethical concerns, must be considered and addressed to ensure the responsible and safe application of mutagenic agents in agriculture.

Gamma ray induced mutagenesis is particularly important in the context of climate-smart agriculture, as it presents opportunities for developing new crop varieties that are better adapted to changing climatic conditions. By creating new genetic variants with desirable traits such as drought tolerance, heat tolerance, and resistance to pests and diseases, gamma ray induced mutagenesis can help to address global food security challenges and contribute to sustainable agriculture practices.

The future of gamma ray-induced mutagenesis in crop improvement and climate-smart agriculture appears promising. Ongoing advancements in radiation technology, high-throughput screening, molecular marker technologies, and gene editing hold great potential for enhancing the effectiveness of gamma ray-induced mutagenesis in crop improvement efforts. These advancements can contribute to the development of new genetic variants with desirable traits.

Moreover, the integration of gamma ray-induced mutagenesis with other techniques, such as gene editing and marker-assisted selection, offers even greater possibilities for plant breeding. By combining these approaches, researchers can leverage the precision and efficiency of gene editing technologies like CRISPR/Cas9 to introduce specific genetic modifications alongside gamma ray-induced mutagenesis. This integration may lead to the creation of new genetic variants with highly desirable traits, such as improved yield, enhanced nutritional content, and enhanced stress tolerance.

However, future research endeavors should prioritize the development of new genetic variants that not only adapt well to changing climatic conditions but also exhibit high yield potential and good quality. It is crucial to ensure that crop varieties resulting from gamma ray-induced mutagenesis meet the demands of a changing climate while also addressing food security challenges.

Moreover, the safety of using mutagenic agents in agriculture must always be considered. Adequate safety measures and regulatory frameworks should be in place to safeguard human health and the environment. This includes careful risk assessments to evaluate potential hazards associated with mutagenesis techniques and the implementation of appropriate safeguards to mitigate any adverse effects.

In summary, the future of gamma ray-induced mutagenesis in crop improvement and climate-smart agriculture looks promising. Continued research and advancements in radiation technology, high-throughput screening, molecular marker technologies, and gene editing hold great potential for enhancing the effectiveness of gamma ray-induced mutagenesis. However, it is essential to prioritize the development of new genetic variants that are well-suited to changing climatic conditions, high yielding, and of good quality, while ensuring the safety and sustainability of using mutagenic agents in agriculture.

Conflict of interest

The authors declare no conflict of interest.

Author details

Godswill Ntsomboh Ntsefong[1*], Fokam Paul Ernest[1], Likeng-Li-Ngue Benoit Constant[1], Tabi Mbi Kinsley[2], Zambou Alain Hervé[1], Mafouasson Hortense Noelle[3] and Bell Joseph Martin[1*]

1 Faculty of Science, Department of Plant Biology, University of Yaounde I, Yaounde, Cameroon

2 Department of Crop Production Technology, College of Technology, University of Bamenda, Cameroon

3 Institute of Agricultural Research for Development (IRAD), Yaounde, Cameroon

*Address all correspondence to: ntsomboh@yahoo.fr; josmarbell@yahoo.fr

References

[1] Bhoi A, Yadu B, Chandra J, Keshavkant S. Mutagenesis: A coherent technique to develop biotic stress resistant plants. Plant Stress. 2022;**3**:100053

[2] Hossain MA, Strezov V, Chan KY, Nelson PF. Gamma radiation-induced mutagenesis in crop improvement. Radiation Physics and Chemistry. 2019;**161**:71-79

[3] Ahloowalia BS, Maluszynski M, Nichterlein K. Global impact of mutation-derived varieties. Euphytica. 2004;**135**(2):187-204

[4] Joanna J-C, Thomas HT, Jochen K, Bradley JT, editors. Biotechnologies for Plant Mutation Breeding: Protocols. Switzerland: Springer Nature, Springer International Publishing AG; 2017

[5] Food and Agriculture Organization of the United Nations (FAO). (2013). Climate-Smart Agriculture Sourcebook. Available from: http://www.fao.org/3/i3325e/i3325e00.pdf

[6] Islam MR, Amin MN, Hasanuzzaman M, Billah M. Gamma radiation induced mutation in rice (*Oryza sativa* L.) for yield and yield contributing traits. Bangladesh. Journal of Botany. 2016;**45**(4):839-846

[7] Kumar S, Singh M, Singh V, Kumar R. Gamma irradiation-induced mutagenesis in wheat for improved gluten quality and baking properties. Cereal Chemistry. 2018;**95**(4):567-575

[8] Rocha MVP, de Oliveira LA, Rocha CB, de Oliveira AC. Gamma radiation-induced mutagenesis in soybean: A review. Journal of Crop Science and Biotechnology. 2020;**23**(3):173-182

[9] Bhatti MA, Iqbal A, Ahmad S, Hussain S, Hussain N. Gamma irradiation-induced genetic variability for drought tolerance in wheat (*Triticum aestivum* L.). Journal of Radiation Research and Applied Sciences. 2020;**13**(3):243-250

[10] Kumar S, Singh V, Singh M. Gamma irradiation-induced mutagenesis in tomato (*Solanum lycopersicum* L.) for heat stress tolerance and fruit quality improvement. Journal of Plant Growth Regulation. 2019;**38**(4):1183-1194

[11] Furbank RT, Tester M. Phenomics - Technologies to relieve the phenotyping bottleneck. Trends in Plant Science. 2011;**16**(12):635-644

[12] Ntsomboh-Ntsefong G, Essubalew GS, Tabi MK, Fentanesh CK, Mahbou Somo Toukam G, Mohammad AS, et al. CRISPR-Cas-based genome editing for crop improvement: Progress, challenges and future prospects. Global Journal of Botanical Science. 2023;**11**:1-6

Chapter 3

Gamma-Ray Spectrometry in Radioactive Prospecting: Application Tool as Detecting Fault Trace

Imaizumi Masayuki

Abstract

In order to clarify the possibility of fault detection by gamma-ray spectrometry (GRS), radioactivity prospecting including soil radon gas survey, etc., was carried out on already-known faults of four areas included active faults. As a result, it was shown that it is possible to detect fault traces with the following three indices: (1) the continuity of anomaly points with an increasing rate Bi/Tl above the threshold defined for each study area. (2) The continuity of peak points of non-diffusive radon. (3) The continuity of conversion points where the average value changes in the stepped fluctuation pattern of nuclide and nuclide ratio. The index of (1) was near all four faults within 0–30 m of the known fault location. All fault-related anomalies of the index (1) were formed by local maxima in Bi and local minima or decreasing trends of Tl. Therefore, it is difficult to detect faults using only total gamma ray measuring equipment such as survey meters, which has been done so far. In order to continue to develop analysis technology for GRS, in addition to accurate GRS measurements, it is also necessary to develop analysis technology using Artificial Intelligence (AI) technology, etc.

Keywords: buried fault, active and non-active faults, soil radon gas survey, increasing rate, non-diffusive radon, stepped fluctuation pattern

1. Introduction

Development of the radioactive prospecting in the early stage was closely related to the uranium deposit development. After World War II, the nuclear development competition (nuclear arms race) intensified. A lot of uranium (U) deposits were discovered by the radioactive prospecting, etc. In the United States of America, although only a little uranium vein was known until 1948, 525 U deposits were discovered in 1953. Although there was only one U mine in Canada in 1952, Canada came to account for about 30% of the world U production in 1959 [1]. The former Soviet Union established the government-run U development enterprise in the Czech Republic, Germany, and Poland. The center of uranium development was Erzgebirge and the circumference near the border between the Czech Republic and Germany. U deposits in 164 places were discovered from 1946 to 1990 [2]. The radioactive prospecting played an important role in these uranium developments. The radioactive prospecting is divided into soil radon (Rn) gas survey that measures α-ray from soil radon gas and gamma-ray spectrometry (GRS) that measures gamma-ray.

The terminology of GRS differs as follows depending on the different methods of acquiring gamma-ray spectra: Air-borne is measured by mounting a measuring device on an aircraft or helicopter, Car-borne is measured by mounting a device on a car, Man-borne (portable measurement) is measured by a person on the ground with a portable detector, etc.

Gamma-ray spectrometry could not discriminate radionuclides from the 1940s to the 1950s, but since the mid-1960s, the demand for monitoring the effects of nuclear tests and advances in computer technology have made field measurement of potassium (K), U, and thorium (Th) possible. The International Atomic Energy Agency (IAEA) began studies in the 1970s to standardize this method. It led to the publication of the first standard technical guide in 1991 [3]. In 2003, the IAEA published a second guideline that added the gamma-ray spectrometry theory to consider the use of GRS for environmental surveys other than for geological survey purposes [4]. This completes the technical system. GRS has already become a mature measurement method with measuring instruments and standardized analytical methodologies [5]. One of the remaining issues is the development of utilization technology for using the U, Th, and K concentration data measured by GRS.

Suran [2] examined which radiological exploration method contributed to the discovery of the 164 U deposits discovered in the Czech Republic between 1946 and 1990. He found that the most effective method was soil Rn gas survey (44%), while air-borne survey (3%) and car-borne survey (9%) were less effective. Therefore, the results of Suran [2] show that it is difficult for air-borne and car-borne surveys to directly identify uranium mines as uranium anomalies.

The reason why GRS, especially air-borne, has regained attention in the field of exploration is that GRS has found not only U and Th deposits but also many metal deposits in the altered zone defined as the high K and low Th/K ratio (e.g., [6]). This is the first turning point in the development of technology for using GRS. This turning point occurred due to the understanding of radionuclide behavior in hydrothermal and alteration/weathering processes. Since this turning point, GRS has been accepted as a geological survey technique for mapping wide-area radionuclide concentration zones, rather than as a technique for exploring the location of anomalous radiation spots.

Discontinuous planes of all sizes, such as faults and joints, are present in rock masses. In particular, research about faults is of special interest to researchers in scientific and engineering disciplines. Because a fracture zone has extensive shearing, the strength and modulus of elasticity of a fracture zone are lower and the hydraulic conductivity is higher than that of the surrounding area. Consequently, assessment of the fracture zone's properties is considered important to the selection of a site, stability analysis, and design of construction works.

Not only for researchers related to earthquake hazards, but also for the general public, distinguishing between "inactive" and "active" faults is one of the most important issues because of the seismic hazard commonly associated with fault activity. In general, scientists define active faults as follows: (1) evidence of historical and/or instrumental seismic activity and/or (2) stratigraphic/morphologic displacements within a time period which can span from the Holocene (10,000 years) to the entire Quaternary. Moreover, it is important to remember that fault slip behavior is often complicated by the mixture of two endmember styles: stick slip (seismic fast slip) and stable sliding (continuous slow creep). These styles can occur in different sections of the same fault and occur in the same section at different times. While the first tends to rejuvenate and keep the fracture network open near the fault, the second allows for fault self-shielding to occur due to the precipitation of fluid constituents flowing through the fracture [7].

Fault trace survey is a major theme of radioactive prospecting. Ambron [8] was the first to find the relationship between faults covered with sediments (buried faults) and radioactivity through α-ray surveys using an ionization chamber. He suggested that the peak positions of α-ray survey can be applied to fault trace survey. Lane and Bennett [9] reported that radon concentrations in groundwater are indicative of known faults. However, it was difficult to investigate a wide area of faults of interest with the α-ray analysis technology at that time. Therefore, Ochiai [10] developed a vehicle-mounted scintillation counter that can continuously measure γ-rays as a Radon index on the ground based on the assumption of the radiation equilibrium of uranium series nuclides. He estimated locations on the faults for bedrock groundwater extraction wells from sequences of γ-ray anomalies.

After his research, in Japan γ-ray prospecting with a survey meter was frequently used for detecting fault traces. This is in contrast to the general use of γ-ray radioactivity prospecting research in the world for the purpose of mineral exploration. On the other hand, some studies in Japan denied the effectiveness of this method for fault survey from some cases where γ-rays did not increase on the fault (e.g., [11, 12]). Imaizumi et al. [12] suggested that the contribution of Bi from soil pore Rn gas to the γ-ray count was about 1.5% based on model calculations, and that it is difficult to estimate the soil Rn gas concentration from γ-rays.

In the field of oil geophysical prospecting, γ-ray (GR) logging was introduced in the late 1930s as the first non-electrical logging method [13]. GR logging is useful in distinguishing between clean sand and shale formations. Then in the 1950s, spectral γ-ray (SGR) logging that could discriminate between nuclides was developed. An initial successful application was the detection of radioactive fractures in Austin Chalk wells, Texas, USA [14]. The Austin formation was originally interpreted as a shale zone. SGR logging was able to distinguish between shale (moderate potassium (K) and thorium (Th) content but low uranium (U)) and U-bearing fracture zones (low K and Th but high U). Based on this information, recompletion of wells increased production sevenfold in some cases [14].

Kimura [15] introduced γ-ray spectrometry for measuring three nuclides into fault surveys and proposed using the Bi/Tl or Bi/K ratio (hereafter, the ratio of three nuclides) as a fault index. He developed a car-borne survey system that detects faults from the increasing rate that is the ratio of the measurement point to the moving average around the measurement point. Imaizumi [16] verified the effectiveness of Kimura's increasing rate on various faults by combining Kimura's system [15] with other surveys such as soil radon gas surveys. The results confirmed the effectiveness of the increase in rate [12, 16–18].

Since 2000, thanks to the development of low-cost, high-precision Si semiconductor detectors for alpha-ray detection and advances in geostatistical analysis technology, significant increases in radon carrier gas (e.g., CO_2) and radon gas were found along active faults (e.g., [7]). In other words, the effectiveness of the alpha-ray exploration method as a tool for fault trace survey has been established. Moreover, numerous radon gas data have advanced our understanding of the structural properties of faults (echelon faults, open cracks, fault clay-sealed cracks, etc.). According to these studies, radon anomalies along fault lines can be continuous or intermittent [7].

On the other hand, regarding γ-ray fault surveys, the relationship between γ-rays and faults remains ambiguous. For example, Szabó et al. [19] reported that faults can be detected using geostatistical spatial structure analysis even with sparse gamma-ray dose rate data (average 3 points/100 km^2), while Jolie et al. [20] suggested that gamma-rays are affected by man-made structures other than faults, so this survey method is not necessarily effective. Yoshimura and Matsumoto [21] showed various examples of the fact that distribution patterns of anomalies depend on the form of the fracture zone (fault clay or open crack, etc.). Some cases even showed that the dose rate decreased over faults. Therefore, at present, even the assumptions of the γ-ray fault surveys are suspicious.

The reason why the radiation dose increases on the fault is not only the radon gas rising along the fault crack but also the concentration of radon parent nuclide elements U and Ra in the fault clay. Therefore, it is impossible to uniquely determine the distribution pattern of γ-ray nuclides, which is an indicator of faults, unless the physicochemical conditions of the faults are elucidated. One way to overcome these problems is to classify and catalog behavioral patterns of gamma-ray-emitting radionuclides (^{40}K, ^{214}Bi, and ^{208}Tl) and alpha-rays (soil radon gas) along survey lines crossing faults of various geological settings and activities (active and inactive faults). We can extract fluctuating patterns of γ-ray nuclides as common fault indices based on the data. However, few studies have clarified the fluctuation behavior of γ-rays on faults caused by U leaching and adsorption/enrichment on clay.

In this paper, Section 2 summarizes the behavior of γ-ray nuclides during weathering and faulting processes, and describes the results of RGS and soil gas radon surveys of exposed faults where the soil has been stripped away. Based on the results, indices for fault detection, such as the Bi/Tl ratio, are proposed. Section 3 presents surveys of four known faults using the proposed index. Section 4 presents the results of regional fracture system surveys, geological surveys, and weathering surveys of the Atera fault area and the southern Abukuma area by a car-borne survey. The original data for Sections 3 and 4 are taken from Imaizumi [16] and Imaizumi et al. [12, 17, 18]. However, these data have been reviewed using Geographic Information System (GIS) techniques. Finally, the study results are summarized and discussed in Section 5, and future studies are recommended based on the indices established by this study.

2. Environmental γ-ray radionuclides

Gamma-ray spectrometry (GRS) is already a mature technology, and its principles and analysis methods have been compiled by the IAEA [3, 4]. Here, only the basics of using GRS as a fault detection method are described. See the IAEA for other basics.

2.1 Natural γ-ray radionuclides

Radioactivity is a phenomenon that changes into a stable nucleus when atomic nucleus with unstable balance of proton and neutron emits elementary particles of α particles (radiation), β particles, and γ photons (ray) over time. Elements with this feature are called radionuclides. γ-rays are electromagnetic waves of surplus energy emitted when an unstable excited nucleus changes to a new stable nucleus. When γ-rays pass through matter, they interact with the electrons and nuclei of matter atoms through phenomena such as photoelectron effects, Compton scattering, and electron pair generation. Under the practical conditions of GRS, Compton scattering is the dominant process. Gamma-ray radiation from soil is predominantly attenuated by water, soil, and organic matter [22].

The most of natural radionuclides that were produced during the formation of the Earth and concentrated in the crust turned into stable nuclides after 4.5 billion years. Currently, there are 17 nuclides with a half-life of more than 700 million years. Among the existing nuclides, the only nuclides that have high radioactivity and are subject to γ-ray spectrometry are ^{40}K, ^{238}U, and ^{232}Th. In all, 0.012% of potassium (K) is ^{40}K, 83.3% of which is β-ray-emitting nuclide, and 10.7% is decayed to ^{40}Ar by EC (electron capture), and 1.46 MeV γ-ray is emitted in the process. ^{238}U and ^{232}Th decay to produce other nuclides (daughter nuclides), and the daughter nuclides undergo further radioactive decay to form a radioactive decay chain (**Figure 1**). Only ^{214}Bi and ^{208}Tl are the nuclides that emit γ-rays that are the target of γ-ray spectrometry (Because the counting rate of γ-rays of ^{214}Pb and ^{228}Ac (actinium 228) is low, they are not the target nuclides of γ-ray spectrometry described below).

Since the half-life of each series of parent nuclides is significantly longer than that of daughter nuclides, the radioactivity of each nuclide in the series is closed state in secular equilibrium, which is the same as that of the parent nuclide in a closed system. In secular equilibrium, the radioactivity of all nuclides in the series is equal, so the nuclide concentration at one stage of the decay chain can be estimated from all daughter nuclide concentrations. Since ^{40}K exists in a fixed ratio to the non-radioactive K isotope, K (%) can be analyzed directly from ^{40}K γ-rays of GRS. The U concentration of GRS is estimated from ^{214}Biγ rays of the daughter nuclide of 238 U.

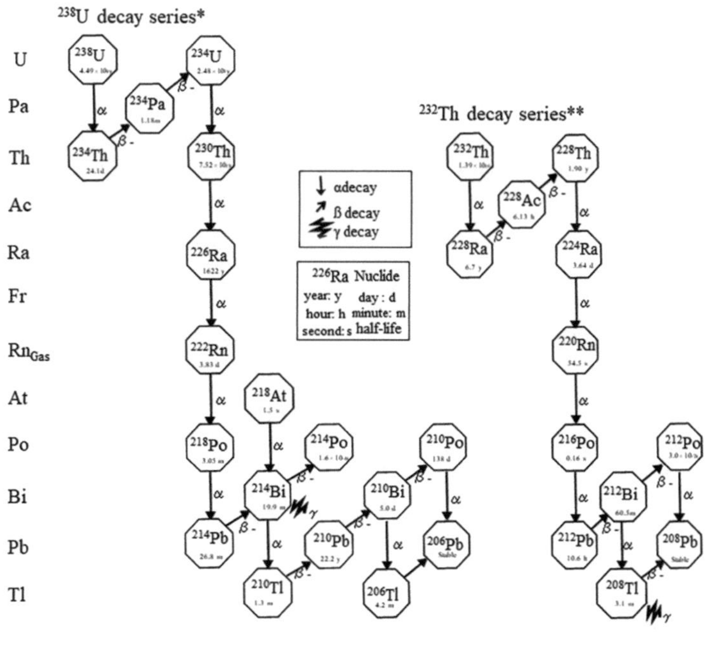

Figure 1.
*Radioactive decay series of ^{238}U and ^{232}Th. Figures from Wikipedia decay chain uranium series and thorium series are edited and added. *https://en.wikipedia.org/wiki/Decay_chain, **https://commons.wikimedia.org/wiki/File:Decay_Chain_of_Thorium.svg*

Therefore, the uranium concentration by GRS is an indirectly estimated concentration. Similarly, Th concentration is estimated indirectly from ^{208}Tl γ rays of the daughter nuclide of ^{232}Th [4]. When one or more decay products in the decay chain are completely or partially removed or added to the system, the decay chain becomes non-equilibrium state. ^{40}K has nothing to do with the non-equilibrium problem. In the Th series, non-equilibrium state rarely occurs due to the low mobility of daughter nuclides. However, in the U series, non-equilibrium states inevitably occur due to selective leaching of decay products (e.g., ^{226}Ra), diffusion of ^{222}Rn gas from soil, and dissolution of ^{226}Ra in groundwater [6].

2.2 Geochemical behavior of radionuclides

2.2.1 Behavior of K, U, and Th in the igneous process

The concentrations of K, U, and Th in igneous rocks vary systematically from ultramafic to felsic. These elements have large ionic radii and are called incompatible elements because they do not fit easily into the crystal lattice of minerals. Therefore, they stay in the magma until the final stage of crystallization differentiation. In igneous rocks, the higher the SiO_2 concentration, the higher the radionuclide concentration in the rock [23]. In acidic rocks such as granite and rhyolite (SiO_2: more than 66%), U is found in the main minerals like feldspar and mica, as well as in accessory minerals (zircon, monazite, etc.) or as a tetravalent element in the interstitial space of the crystals. Th also occurs not only as a tetravalent element in a separate mineral, but also by replacing U, Zr, rare earth elements, etc., in the mineral [23].

2.2.2 Behavior of U in intrusive rock emplacement process

Intrusive rock is formed when plutonic rocks such as granite and diorite intrude and solidify into the crust as batholiths and dykes. It is well known that rocks that touch granite intrusive rock undergo contact metamorphism and contact metasomatism because of the intrusions. Analysis of the batholith's drill core shows that the closer to the surface layer, the stronger the radioactivity. In other words, radioactive materials are more concentrated near the roof pendant (**Figure 2(a)**).

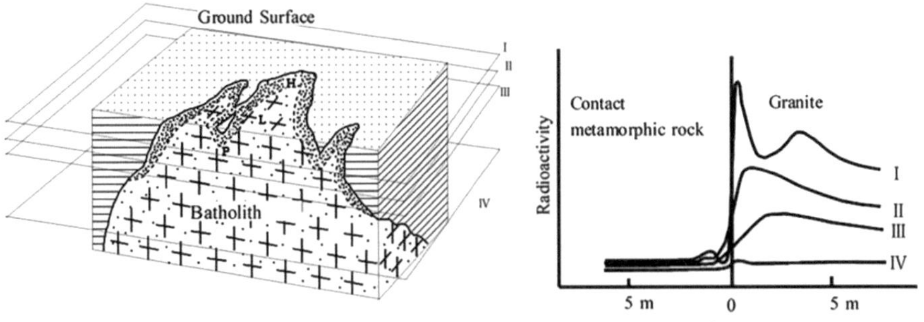

Figure 2.
Radioactivity distribution in granitic intrusive rocks (a) and radioactivity distribution near the interface of contact metamorphism (b) (edited from figures in Hatsuda [24]). P: Roof pendant, H: High activity part, and L: Low activity part. The density of dots represents the density of radioactive elements.

Uranium remains in magma until the final stage of magma crystallization differentiation. As it consolidates into crystals at the terminal stage, a relatively distinct radioactive anomaly forms at the upper rock boundary of the granite body. Under the same conditions during contact metamorphism, the curves of radioactivity intensity distribution across the contact area of the intrusive rocks show similar patterns. These patterns in Japan are divided into four types, as shown in **Figure 2(b)** [24]. Type I is a pattern that appears during typical contact metamorphism. It is thought that this type of intrusive rock boundary can be detected as an anomaly in radiological surveys. An example of this is shown in the car-borne survey of the Atera fault area in Section 4. The pattern of radioactive anomaly becomes weaker from type II to type IV. Hatsuda [24] interpreted that type IV indicates the radioactivity distribution in the deep part of the intrusion where granitization was remarkable. An example of this is shown in car-borne surveys of the Abukuma area of Section 4.

2.2.3 Behavior of K, U, and Th in the sedimentary process

Ordinary sedimentary rocks always contain U and Th, although there are differences depending on the region and material. U is correlated with C in sedimentary rocks and is relatively well correlated with K in Neogene mudstones. Humus and other organic matter adsorb U. U is also adsorbed by iron hydroxide and clay minerals. U is more abundant in rocks like black shales that are deposited in strongly reducing sedimentary environments with a lot of organic matter and mainly clay minerals. U is also concentrated in phosphorite. Therefore, the distribution of U concentration in sedimentary rocks is determined by the paleogeography of sedimentary basins [25].

2.2.4 Behavior of K, U, and Th in the weathering process

When rocks underground are exposed to the temperature and pressure above the ground, they weather and decompose, changing their composition. Rock components dissolve and recrystallize to form clay minerals, turning the rock into a parent material for soil. Moreover, when organisms and organic matter interact with the parent material, soil is formed. The concentration of radionuclides in soil mainly depends on the geology of the bedrock, the formation of clay minerals by weathering, and the geochemical behavior of radionuclides [6]. Generally, the composition of radionuclide concentrations in the parent material is inherited by the soil, but some are changed through the pedogenesis process. General trends in soil concentrations of K and U, Th series nuclides show a positive correlation as well as rock trends.

Potassium (K) is rich in felsic rocks because it is found in main minerals like feldspar and mica. It is rare in basic and ultrabasic rocks. It is not found in carbonate rocks at all [22]. Feldspar and mica are present in the sand fraction in the early stages of soil formation. But they recrystallize into clay minerals and disappear because of weathering. The K in the feldspar crystal lattice is usually washed away by rainwater when the feldspar changes into kaolinite with almost no K. When mica changes into illite at the first weathering stage, some K stays in the illite crystal lattice. At more weathering, illite changes into low K-concentration clay minerals (vermiculite and smectite). At the final weathering stage, low K-concentration clay minerals change into kaolinite [26].

Uranium and Th have similar chemical behavior in magma. In unweathered igneous rocks, both elements are tetravalent. But in the process of weathering, they show a completely different chemical behavior. This is because of the difference in sensitivity to oxidation and reduction changes. In the oxidized atmosphere, U becomes hexavalent uranyl ion (UO_2^{2+}) and forms many kinds of complex ions. Since these complex ions have relatively high solubility, hexavalent uranium easily moves with groundwater. But uranium in a water solution is adsorbed to an adsorbent such as clay minerals, organic matter, trivalent iron hydroxides, and zeolites. Montmorillonite has more uranium exchange capacity than other clay minerals. When hexavalent uranium is reduced for some reason, it becomes tetravalent and precipitates as poorly soluble uranium dioxide (UO_2), and recrystallizes as a secondary mineral. So, U is sensitive to oxidation and reduction changes and is mobile [26]. On the other hand, Th keeps its tetravalent state, so its solubility and mobility are low [26]. Even after K and U are removed by weathering, Th remains as a residue and tends to keep the source rock composition. Note that in the analysis of GRS data, this Th characteristic, specifically the Th nuclide ratio, can be used to analyze the migration and enrichment of U.

2.2.5 K/Th and U/Th ratios as the indicator of the sedimentary environment

Based on the difference in how U and Th react to oxidation and reduction changes, the Th/U ratio (hereafter, the "ratio" is omitted) is used as an indicator of the paleo-redox environment during deposition, and Th/K is used as an indicator of clay mineral types.

2.2.5.1 Th/U

Adams and Weaver [27] concluded that Th/U ratios are often strongly related to the sedimentary environment based on laboratory analysis of many samples. So far, studies have shown a strict relationship between the Th/U and the sedimentary environment. This relationship shows that high Th/U is typical of the continental environment and low Th/U of marine settings. The boundary value of this ratio was determined as follows: $Th/U > 7$ for continental deposits, $Th/U < 7$ for marine deposits, and $Th/U < 2$ for high chance of reducing conditions. $Th/U > 7$ for high chance of oxidation conditions [28].

2.2.5.2 K/Th

Schlumberger [29] proposed a reference thorium-potassium cross-plot that distinguishes five typical minerals that are important for the oil industry: chlorite, glauconite, illite, kaolinite, and smectite by the Th/K ratio, based on many data analysis results (**Figure 3**). In this figure, plagioclase (Th/K = 0.3–0.6) has the lowest Th/K ratio. The Th/K ratio increases in the order of mica, illite, mixed layer clay, and montmorillonite. Kaolinite as the final weathered mineral has Th/K = 12–25.

Many researchers have used them to identify types of clay minerals. Nuțu-Dragomir et al. [30] used an improved Schlumberger Th-K cross-plot for fault survey in shale and marl regions. Their Th-K cross-plot (**Figure 3**) showed that the clay minerals in the fractured zone were montmorillonite and the clay minerals in the outcrops of non-fractured zone were illite-smectite mixed layer minerals. Stahr et al. [26] applied the cross-plot for the discrimination of soil types of Acrisols (cation exchange capacity (CEC) <24 cmol. kgL^{-1}) and Alisols (CEC ≥ 24 cmol. kgL^{-1}) that require laboratory measurements in northern Thailand. Th/K = 16 by GRS in the field was able to distinguish between Acrisols and Alisols.

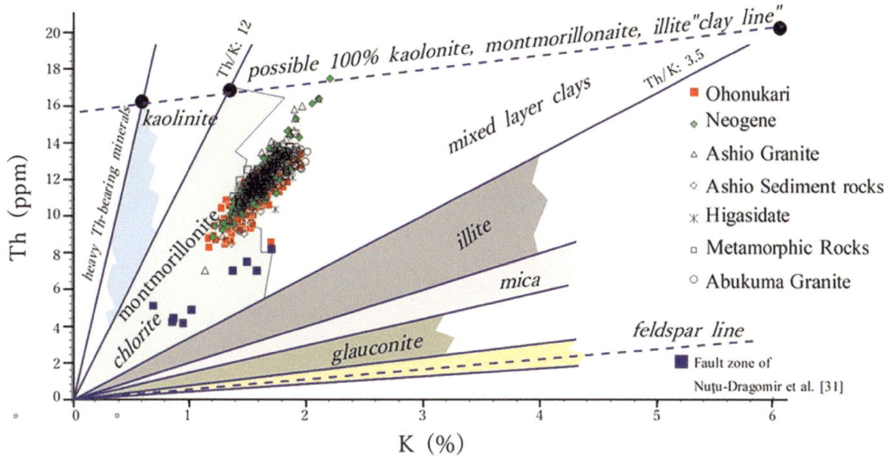

Figure 3.
Thorium-potassium cross-plot for mineral identification. The Th-K cross-plot is modified from Schlumberger [29] by Nuţu-Dragomir et al. [30]. The data plotted are for each rock body in the Tanagura fracture zone in the southern Abukuma region. See text in section 4–2 for details.

Th(ppm)/K (%)	Minerals
<0.6	Feldspar
0.6–1.5	Glauconite
1.5–2.0	Mica
2.0–3.5	Illite
>3.5	Mixed-layer clays
> = 10	Kaolinite and chlorite

Table 1.
The relationship between K/Th and clay minerals [28].

Using cross-plot diagrams to determine the types of clay minerals has some drawbacks, because the interpretation results cannot be shown as a function of depth. Showing the Th/K as a function of depth allows us to track the changes in the types of clay minerals in the stratigraphic column. In this case, the relationship between K/Th and clay minerals is classified by the K/Th value in **Table 1**.

2.2.6 Behavior of K, U, and Th around the fault fracture zones

2.2.6.1 Structure of the fault fracture zones

A fault is shown as a geometrical plane on geological maps. But an actual fault has the following forms: The fault gouge (fault clay) and the fault breccia (both together called the intrafault materials) that are formed by the crushing and friction of fault movement are layered between two fault planes that run almost parallel [31].

When a large-scale shear fracture occurs, the breccia and clay mix because the rock mass is broken in a relatively wide width. This state, made of clay (fault gouge) and relatively coarse-grain fault breccia, is called a fault fracture zone. In the fault fracture zone, the rock mass, which usually has many cracks, lies beyond a certain constant

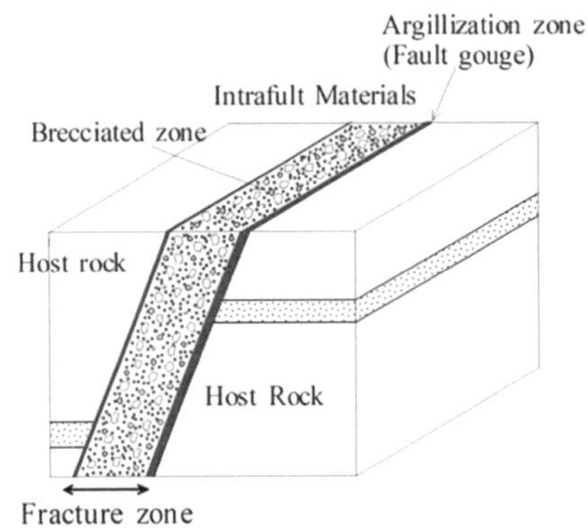

Figure 4.
Schematic figure of fault fracture zones (modified from Tanaka [31]).

width of breccia and clay. The cracks gradually decrease and turn into normal rock at a certain distance from the center of the fracture zone.

A clay mineral in fault gouge is usually formed by chemical weathering and alteration due to reaction with groundwater after crushing [31]. The texture of the fault fracture zones seen in outcrops shows a banding structure, with a zone mainly made of fault gouge (clay zone) and a zone of fault breccia (brecciated zone) running parallel to the fault. Generally, the clay zone is located on both sides of the fracture zone (**Figure 4**) [31]. Therefore, U tends to be enriched in clays at both ends of the fracture zone.

2.2.6.2 Geochemistry of the fault

The formation of the clay minerals in the fault fracture zone is strongly influenced by the type of host rock (original materials) and the surrounding temperature and pressure, as well as the generation of clay minerals by general weathering. The composition of the clay minerals in the Atotugawa fault fracture zone in Japan showed that feldspar and quartz, etc., which are main minerals in the host rock (granite), are hardly found, but montmorillonite has been newly formed [32]. The amount of montmorillonite, compared to the whole mineral composition, is high in the argillization zone, and low in the brecciated zone.

The non-equilibrium data of uranium series show that U very slowly leaches from the primary mineral [33]. On the other hand, the crack in the fault is often characterized by various filling and alteration materials. The most common materials are hematite and goethite. It is often observed that uranium is concentrated around the secondary Fe^{+3} mineral. Smellie et al. [34] showed a Fe-U-Ra vertical profile in the crack of faults in three different granite regions, where total Fe, Fe^{3+}, and U generally increased toward the crack, although the Th concentration was almost constant.

2.2.6.3 U/K and Th/U curves as indicators for crack exploration

The principle of detecting fracture systems in carbonate formations by spectral γ-ray (SGR) logging is based on the fact that under the reducing conditions in which carbonate rocks form, hydrothermal circulation can cause uranium salts to precipitate within the fractures [28]. However, in complex crack systems, low concentrations of

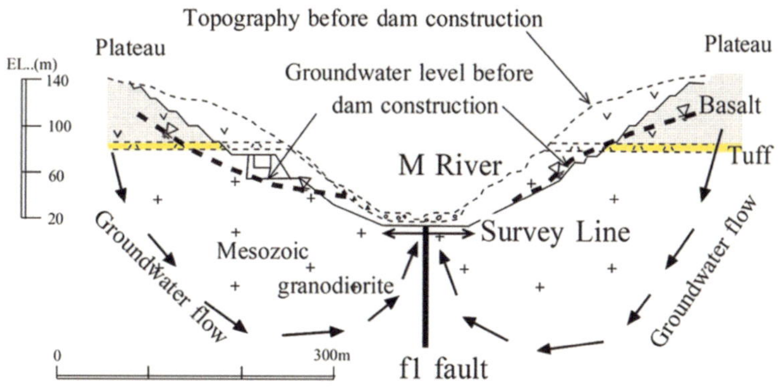

Figure 5.
Geological section of the valley across the M River at the F dam site.

radioactive elements can obscure the crack locations in the uranium curve of the SGR curve. Therefore, plotting the ratio curves, such as U/K, Th/U, and Th/K, helps to better identify the crack locations. Of these ratio curves, matching the U/K and Th/U curves is highly recommended. These curves are inversely proportional, with changes in uranium content plotting the curves in opposite directions [28].

2.3 Indices for fault trace detection for radioactivity prospecting

Buried faults are faults that are covered by newer sediment or soil and cannot be seen directly. The planned inspection gallery site of a dam with a fault is a valuable example where the buried fault can be seen because the surface sediment has been removed. The planned site for F Dam was one of these examples.

Figure 5 shows a geological section of the valley across the M River at the F dam site. The width of the M river is about 20 m. The slope on both sides of the valley is about 30°. About 140 m above the slope there is a plateau. The geology of the plateau consists of Neogene basalts in the upper part and Mesozoic granodiorite in the lower part. The dashed line in the figure shows the terrain before the dam construction. At the time of the survey, river bed gravel, river terrace deposits, and weathered parts of the bedrock had been taken away to expose fresh bedrock, so F1 faults could be seen (**Figure 6**).

The fracture occurrence of bedrock is explained by Japanese rock mass classification system. Rock mass classification in Japan classifies rock mass into four ranks from A to D. C ranks are subdivided into C_H (high), C_M (middle), and C_L (low). D ranks are subdivided into D_H and D_L. A and B indicate hard rock with few cracks, and $C_H \sim C_L$ indicate medium hard rock with cracks. Crack density increases from C_H to C_L. D_H is very soft, has no cohesion between joints, and falls apart with a slight hammer blow. Clay is on the peeled surface. D_L is a clayey state (e.g., [35]).

Figure 6 shows the bedrock classification map and the distribution of springs at the planned inspection gallery construction site. The classification map shows the locations of high-angle f1-f8 faults. Except for the f1 fault, there were small faults with a fracture zone width of 0.3 to 0.5 m. The f1 fault was represented by D_L rock with a width of 0.5 to 2 m. D_H rock with a width of 2 to 5 m was located on the west side. Five meters away from the f1 fault, C_H-C_M class hard rock with few cracks was distributed.

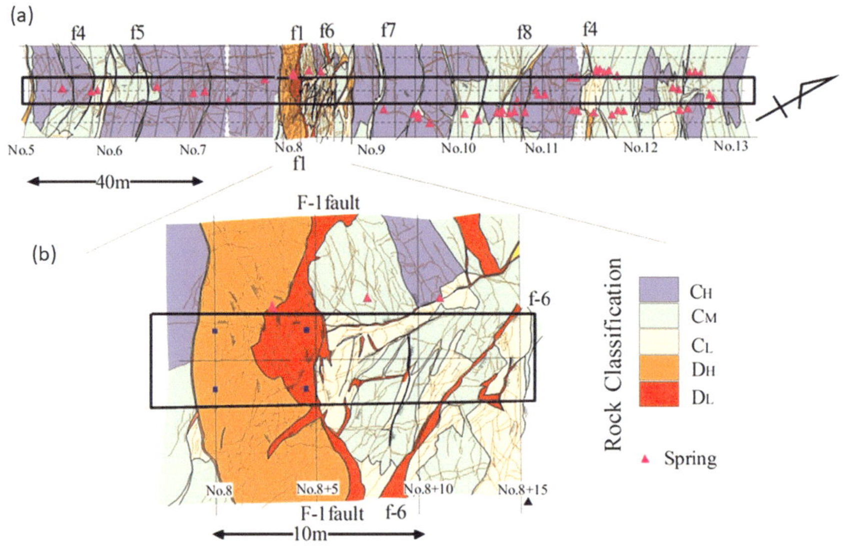

Figure 6.
The bedrock classification map and the distribution of spring water at the planned inspection gallery construction site of the F dam site.

Many springs were distributed in the riverbed. The amount of spring water from each spring flowed at the rate of several liters per minute (liters/min). The spring distribution was mainly located in the cracks of C_H rocks on the east side of the fault and the boundary cracks between C_H and C_M rocks. There was almost no distribution of springs near the f1 fault.

A 50-m-long survey line centered on the f1 fault for radioactivity prospecting was set up. GRS and radon gas concentration were measured at 5 m intervals along the survey line. The GRS sampling time was 300 s. Methods for GRS and radon gas survey are described in Section 3.3.

Figure 7 (a) shows the fluctuation patterns of Bi, Tl, Bi/Tl, and Rn gas concentrations and (b) shows the distribution of fractures (extracted from **Figure 6**). The white areas in the figure show the C_M and C_H rocks with few cracks. Across the main fault zone from f1 to f6, where the density of fractures is the highest, Bi and Bi/Tl drop quickly to minimum values (indicated by a green triangle). On the other hand, Tl increased slightly. Since the pattern of Tl shows the original pattern, the Bi and Bi/Tl patterns may reflect the leaching of U from the host rock, granodiorite.

On both sides of the main fault zone, relatively large peaks in the Bi/Tl ratio were seen. The largest peak is about 10 m from the f1 fault on the east side (shown by the vertex position of orange triangle). In this part, Tl makes a local minimum. Since there is a proportional relationship between Bi and Tl in the original granodiorite, the original Bi should also have made a local minimum. But it is thought that Bi was locally maximized due to the enrichment caused by precipitation of U (indicated by a red triangle). Bi/Tl highlights this phenomenon more than the fluctuation of Bi (the orange triangle is bigger than the red triangle). The Rn gas concentration showed a big peak of about 3 kBq/m^3 near the boundary between the D_H and C_L rocks, which is several meters away from the f1 fault (D_L rock). Note that this radon gas peak near the f1 fault does not affect the Bi fluctuation pattern.

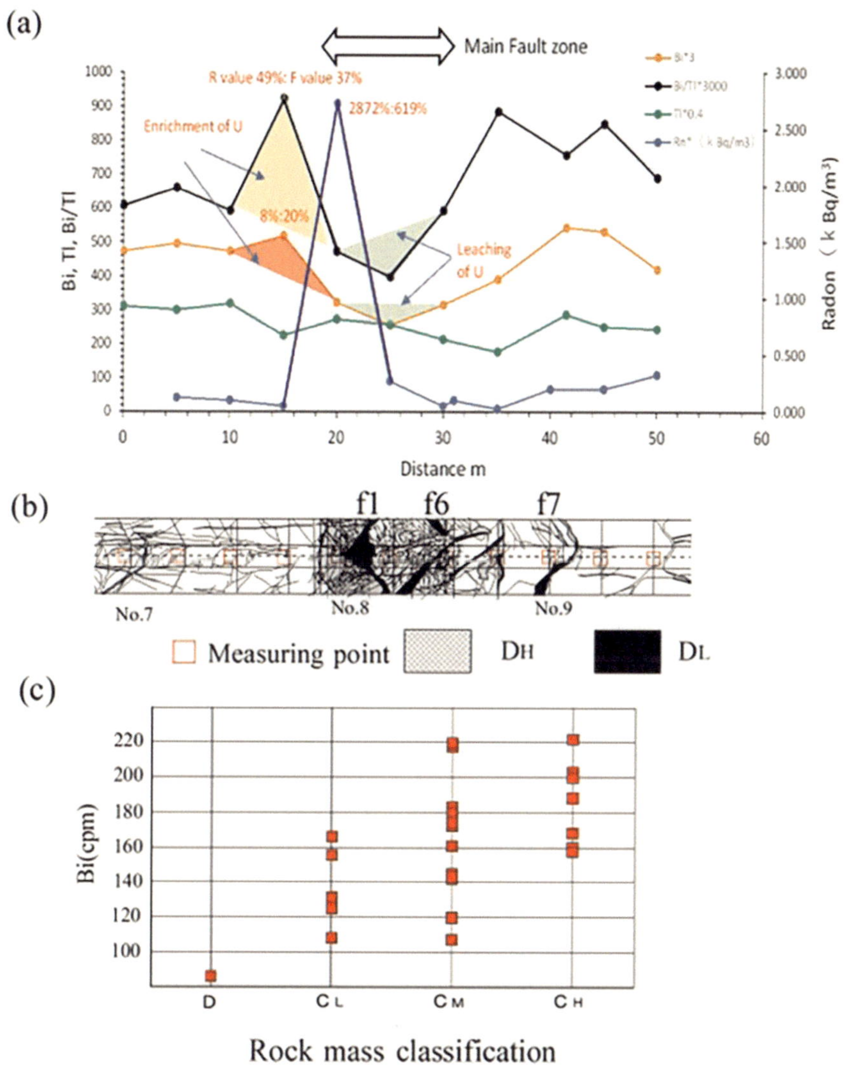

Figure 7.
*Fluctuations of Bi, Tl, Bi/Tl, and Rn gas concentrations (a), the distribution of fractures (b), and the relationship between Bi and rock class classification for the entire excavated surface in the F dam site. In the figures of fluctuation curves below, the values have been processed in order to plot multiple curves in one figure. For example, Bi/Tl*1000 is Bi/Tl multiplied by 1000. See text for red and green triangles.*

Figure 7(c) shows the relationship between Bi and rock class classification for the whole excavated surface. As the bedrock condition gets better (D- > C_H), Bi tends to increase. This phenomenon shows that the leaching of U progresses with the weathering, and the lower the rock class, the lower the U content.

From this figure, the following points can be made about the detection of faults in granitic regions by radioactivity prospecting: (1) The fault fracture zone was the U leaching zone. (2) The locations of the local minimum of Bi and Bi/Tl are indicators of the central location of the fracture zone. (3) The local maximum locations of Bi and Bi/Tl, here the peak on the left side of the f1 fault was about 10 m away from the center of the fracture zone, which is the entrance of the fracture zone. (4) Although the Rn gas peak location is several meters away from the center of the fracture zone, Rn gas is the most sensitive fault indicator. (5) If an error of about 10 m is allowed, the local maxima of Bi/Tl also serve as fault indicators. (6) The leaching zone of U is wider than the enriched zone of U, so the local maximum point of Bi/Tl is better than the local minimum point of Bi/Tl as an index for exploration.

Based on the above observations, in the exploration of known faults in the next section, the local maximum peak of Bi/Tl (anomaly point) and peaks of radon concentration are chosen as indices for fault trace detection.

2.4 Definition of anomaly point for fault detection index

Traditionally, anomaly points have been defined as outliers with a threshold of mean $\pm$ 2 $\times$ standard deviation of a single population assuming a normal distribution in a parametric test, and anomaly points outside this threshold [36]. This method is intended to identify the extreme values of the statistical normal distribution, and geochemical problems often do not warrant statistical assumptions (e.g., [37]). In the analysis of radiological exploration, especially in the analysis of car-borne survey data, a survey line is set across the distribution of multiple background populations (stratum), so it is not possible to define outliers by statistical methods.

Kimura [15] proposed the following equations for increasing rate, specifically assuming the evaluation of anomalies by the car-borne survey (**Figure 8**). Increasing rate: $F_{(f,\ p)}$ is defined as the ratio of the movement average of forward and backward to the measured value given by:

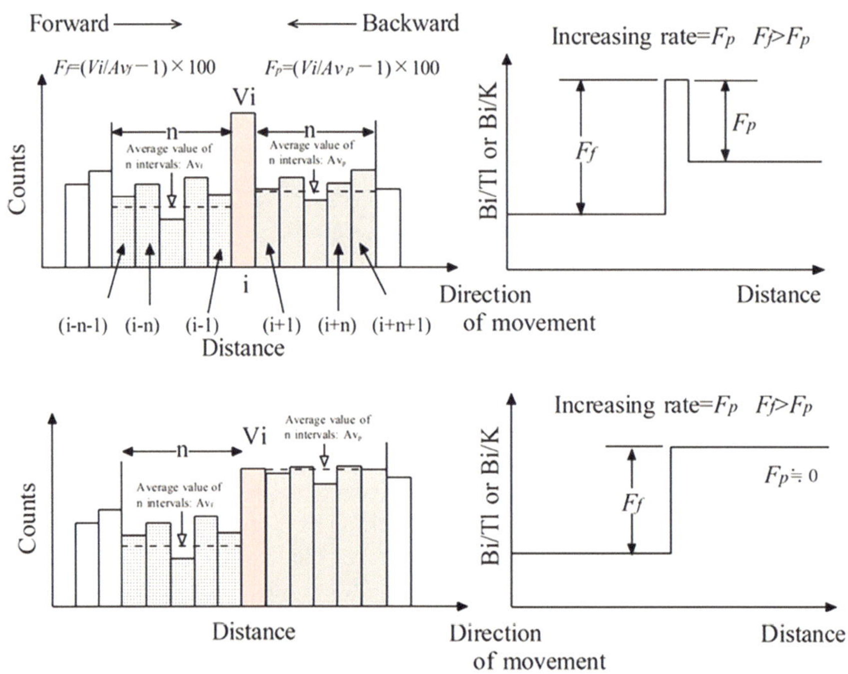

Figure 8.
Schematic diagram explaining the calculation method of radioactive survey anomalies by Kimura [15].

$$F_f = \frac{V_i}{\left(V_{(i-n-1)} + V_{(i-n)} + \cdots\cdots + V_{(i-1)}\right)/n} - 1 \times 100 \quad (1)$$

$$F_p = \frac{V_i}{\left(V_{(i+n-1)} + V_{(i+n)} + \cdots\cdots + V_{(i+1)}\right)/n} - 1 \times 100 \quad (2)$$

where V_i indicates the measurement value at i point. $V_{(i-n-1)}$ and $V_{(i+n+1)}$ indicate the measurement values at the n-front and n-post of i point.

A radioactive anomaly cannot be detected by values of either Ff or Fb alone. But when both are used, the position of the stratum boundary line with different nuclide compositions can be identified as the point of rapid increase of either Ff or Fb. The position of the fault line can be clearly identified as the point of rapid increase of both (**Figure 8**). The increasing rate: R is defined as the smaller of Ff and Fp. Anomalous points are set at points above the threshold based on the coincidence of increasing rate points and fault lines at each field. For example, on the Atera fault, anomalies above the threshold of R = 10% tended to match with the fault location.

Figure 7(a) shows the R-value of Bi, Bi/Tl, and Rn gas concentrations and the rate of increase (F-value) from the average value of the survey line. The F-value for the small Bi peak is 20%, while the R-value decreased to 7.8%. This is due to the distorted peak shape (Ff = 42 and Fb = 7.8). But for Bi/Tl, which produces a clear peak, the F-value was 37%, but the R-value grew to 49%. Similarly, at the Rn gas concentration that produces a distinct peak, the F-value increased by 619%, but the R-value grew to 2872%. This large rate of increase shows that Rn gas is a sensitive fault index. In this way, when the shape of the peak is clear, using the R-value can increase the rate of increase from the simple average value.

3. Fluctuation patterns of nuclide, nuclide ratio, and radon gas concentrations across known faults

3.1 Method

3.1.1 Man-borne survey

The portable pulse height analyzer made by Clearpulse Co., Ltd. and scintillation detector (sodium iodide (NaI) crystal-like cylinder of 13 cm in diameter and height with a photomultiplier tube) were used for the man-borne survey. A detector with energy resolution of 7% or less was used. The NaI detector was placed on the surface of the earth for the measurements. Each measurement time was for a duration of 5 min. The error ($\sqrt{N}$) in the measurement value (N), in this case, was 6% or less ($\sqrt{N}/N$) concerning Bi, whose discharge energy is smaller than those of the other nuclides.

3.1.2 Car-borne survey

The pulse height analyzer (made by Clearpulse Co., Ltd.) with 1024 channels designed for car-borne survey and 18 NaI detectors (the same size as the man-borne detector with resolution 7% or less) were used for the car-borne survey. A stabilizer mechanism that tracks the peak of gamma-rays from a 137Cs source (37 kB) installed near the detector was constantly operating during measurements to keep the amplification rate constant. The detectors of car-borne were placed on one side of the luggage compartment and one side of the detectors were shielded to reduce γ-ray counts from the road pavement under the car. These measures gave direction to the measurement efficiency of the detector [15]. The stripping ratio method [3] was applied to correct the Compton scattering of the γ-ray spectra.

The measurement method was as follows:

1. Running car measurement [15]: The speed of movement was kept at 4 km/h. Sampling time was taken at 30 s intervals. The maximum measurement error ($\sqrt{N}/N$) of Bi was 5% or less. The measurement value indicates an integrated value of radioactivity distribution for about 30 m in length. The location of the measurement point includes an error of about 30 m back, either side, due to various conditions of the measurement.

2. Stopping car measurement: The car was stopped at intervals of 10 m for measurement. The measurement was done three times, every 30 s. The average value of the three measured values was used for analyzing the fault.

3.1.3 Soil radon and CO_2 gas surveys

The soil radon gas survey using the Picorad detection (1) and soil CO_2 gas survey using the detection tube (2) were used in conjunction with the GRS survey.

1. Picorad detection (Packard Co., Ltd.) that is a granular-activated carbon canister method was used for soil radon gas survey. A Picorad detection device consists of a 20-ml plastic vial with cap, including activated carbon. A hole was made with a diameter of 20 cm and 30–70 cm in depth. The Picorad vial, without a cap, was placed inside the bottom of the hole. The container was covered up with soil. The Picorad was left undisturbed for 24 or more hours to adsorb radon surrounding it. A liquid scintillator was injected into the detector after carrying it to the laboratory. The α- and β-rays of the liquid, into which the radon was transferred from the activated carbon, were measured by a liquid scintillation counter. The correction of adsorption and collapse under exposure, and the attenuation of time after collecting and measuring, was executed with the special software offered from Packard Co., Ltd. The limit of detection is 14.8 Bq/m^3.

2. A detection tube made from glass of Kitagawa method (KOMYO RIKAGAKU KOGYO K.K.) was used for the soil CO_2 gas survey. A handheld vacuum gas sampler was used to vent a constant amount of sample gas to the detector tube. The measurement range was 0.1–5.2%.

3.2 Overview of topography and geology for the known fault test sites

3.2.1 Selection of known fault test sites in Japan

Since Japan has a complex geological structure, it is possible to investigate various geological faults. Four test sites with different geological settings shown in **Figure 9** were selected. In this section, first, the relationship between the four test sites and the geological tectonic regions of Japan is described. After that, an overview of the topography and geology of each test site is presented.

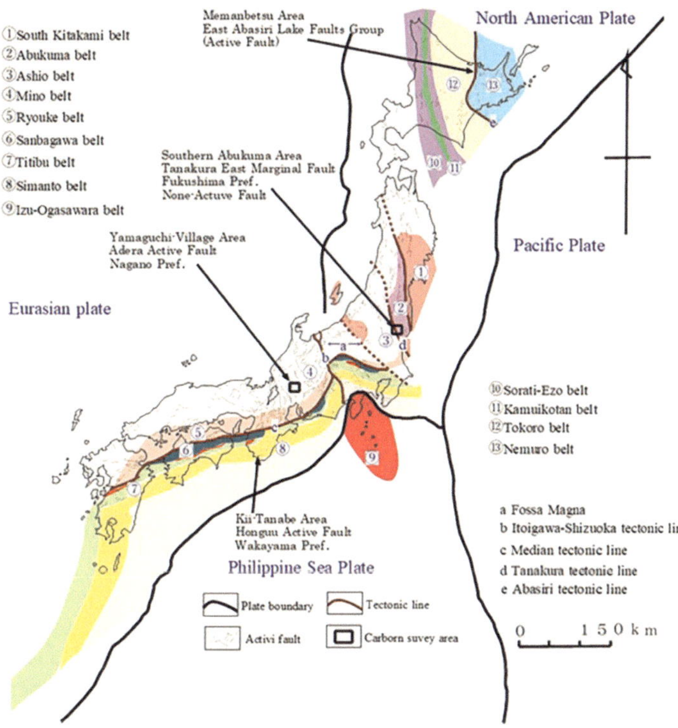

Figure 9.
Location map of four known fault test sites on the geological tectonic region map of Japan. The geological tectonic region map is edited from the original map of the National Institute of advanced industrial science and technology (AIST) (https://www.gsj.jp/geology/geology-japan/geology-japan/index.html), and the active fault lines are edited from the original map of the active fault society [38]. Note that not all geological tectonic regions are shown.

Basically, the subduction in the trench and the related volcanic activity form the backbone of the Japanese island arc. Sedimentary rocks and metamorphic rocks in Japan, especially, are characterized by the continuous zoning of rocks that were formed over a certain period of time. These boundaries are often connected by faults, some of which are called tectonic lines. The geological structure of present-day Japan was formed during the new orogeny due to the development of the curved archipelago after the Miocene. **Figure 9** is a map compiled by the Geological Survey of Japan showing geological regions in Japan that have been proposed so far. This figure shows the differences in rocks underlying the Japan island arc, mainly based on large-scale faults, accretionary prism formation age, and metamorphic age.

Japan's largest island (Honshu) is roughly divided into two parts, northeastern Japan and southwestern Japan, by a rift zone called the Fossa Magna that runs north-south through the center (as indicated by <- a ->) (The tectonic line on its west side is the Itoigawa-Shizuoka tectonic line (b line), and the tectonic line on its east side is not clear. Here, it is indicated by a dotted line). The Fossa Magna has thick deposits of submarine volcanic products and sediments from the mid-Neogene period. They are covered by a north-south chain of volcanoes, including the Quaternary Mt. Fuji.

Cenozoic rocks are widely distributed in northeastern Japan. Many of the Pre-Tertiary rocks are isolated massifs. Granites are distributed in the Abukuma belt (2), and sedimentary rocks and granites are distributed in the Ashio belt (3). Both belts are separated by the Tanagura fracture zone (the d line). On the other hand, in southwestern Japan, Paleozoic and Mesozoic rocks are widely distributed, while Cenozoic rocks are narrowly distributed. Southwestern Japan is further divided by the Median Tectonic Line (the c line) into an inner belt (on the Sea of Japan side) and an outer belt (on the Pacific Ocean side). The following conspicuous zonal structures parallel to the Median Tectonic Line are distributed in the outer belt of southwestern Japan: the Sanbagawa belt (a high-pressure metamorphic belt) (6), the Chichibu belt (a Permian-Triassic accretionary prism: hereafter written as AP) (7), and the Shimanto belt (a Cretaceous-Neogene AP) (8). The Shimanto belt is divided into three sub-belts by two tectonic lines that include the Honguu fault, which are not shown in the figure. The inner belt includes the Ryoke belt (5) (a high-temperature metamorphic belt) and the Mino belt (mainly Jurassic AP) (4). The Mino Belt has several active faults, including the Atera Fault. The eastern part of the Sorachi-Ezo belt (10), which runs north-south through central Hokkaido, is a geological division different from that of the main island of Japan. To the east of the Sorachi-Ezo Belt, the Tokoro Belt (12) (a Cretaceous-Paleogene AP) and the Nemuro Belt (13) (Cretaceous-Cenozoic sedimentary rocks of the Kuril Arc) are distributed. The boundary between the Tokoro belt and the Nemuro belt is the Abashiri Tectonic Line (the e line).

3.2.2 Yamaguchi village area-Atera active faults

The study area is Yamaguchi Village, which is now west part of Nakatsugawa City, Gifu Prefecture, which is located at the southeastern end of the Atera main active fault (**Figure 9**). The Atera fault can be traced for about 67 km. Yamaguchi village is located about 340 km west of Tokyo. The Kiso River flows east-west through the northern edge of the study area and north-south through its west side (**Figure 10**). The Atera fault splits the study area into northeastern and southwestern parts. The Atera fault is the boundary between mountainous area (Mt. Takadoki, with an altitude of 1038 m) and a plateau area (Mt. Bonten, 696 m) in the Mino belt. This geography is caused by a large-scale left slip of the fault so that the NE side of the fault is uplifted relative to the SW side.

The geology of the study area is mainly underlaid by late Cretaceous to Paleogene Nohi rhyolite and granite. The granite caused contact metamorphism during intrusion. The late Miocene to Pliocene Seto Group covers them and is distributed in the Kiso River tributaries. Quaternary sediments are only distributed on a small scale along the Kiso River (**Figure 10(b)** [39]).

The Atera fault is a fracture zone that is composed of parallel faults or echelon faults in a Northwest-Southeast (NW) direction. Faults that are diagonal to the Atera fault are the Magome-Pass fault and its parallel faults. These faults pass from Agatuma Pass to Magome Pass and are cut by the Atera fault [39].

3.2.3 Memanbetsu area-Lake Abashiri east coast active faults

The study area is located in Memanbetsu Town, which is now part of Ōzora Town, in eastern Hokkaido, where active faults run north-south in the eastern coast of Lake Abashiri. The distance from Sapporo to Ōzora Town is 257 km. According to the Active Fault Society [38], the Lake Abashiri east coast active fault group comprises three faults that run parallel to each other in a north-south direction on the east coast of Lake Abashiri (**Figure 11**). The Notori active fault, located on the eastern side of the group, was chosen as the target fault. The survey area has a hilly topography at the northern foot of Mt. Mokoto (with an altitude of 1000 m), which is part of the outer

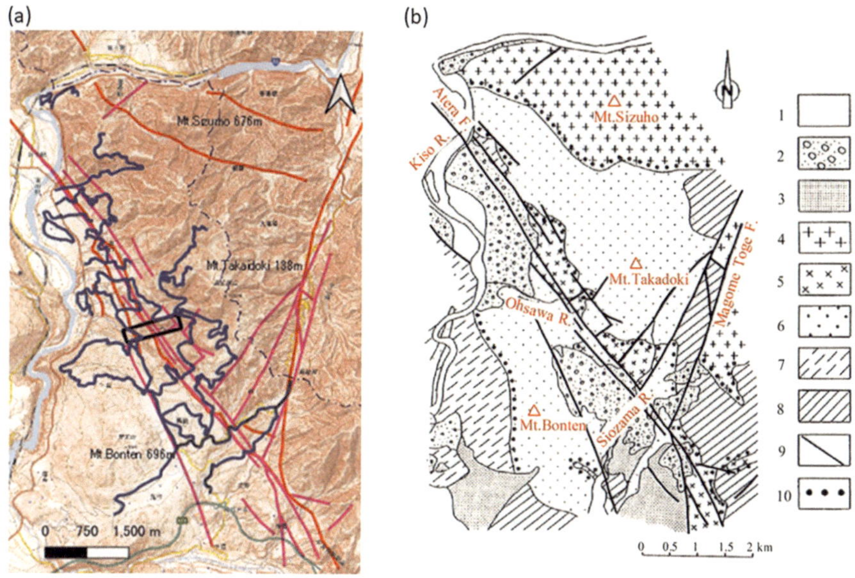

Figure 10.
Topographic map (a) and geological map (b) of the Yamaguchi village area (edited from Yamada [39]) of the study area. Topographic map (a) is displayed with a red relief image map overlaid on the geospatial information Authority of Japan (GSI) tiles. The red relief image map [40] is a topographic map that emphasizes ridges (white) and valleys (red) created by system for automated geoscientific analyses GIS (SAGA-GIS) (https://saga-gis.sourceforge.io/en/index.html). The black frame indicates the man-borne survey area. The blue curves indicate survey route of the car-borne survey. The geological map legends are as follows; 1: Alluvium or talus deposit, 2: Terrace deposit or old alluvial fan deposit, 3: Seto group, 4: Naegi-Agematu granite, 5: Inagawa granite, 6: Granodiorite, 7: Non-classified Nouhi rhyolites, 8: Fujimi-dai welded tuff, 9: Fault, and 10: Strongly radioactive zone by contact metamorphism.

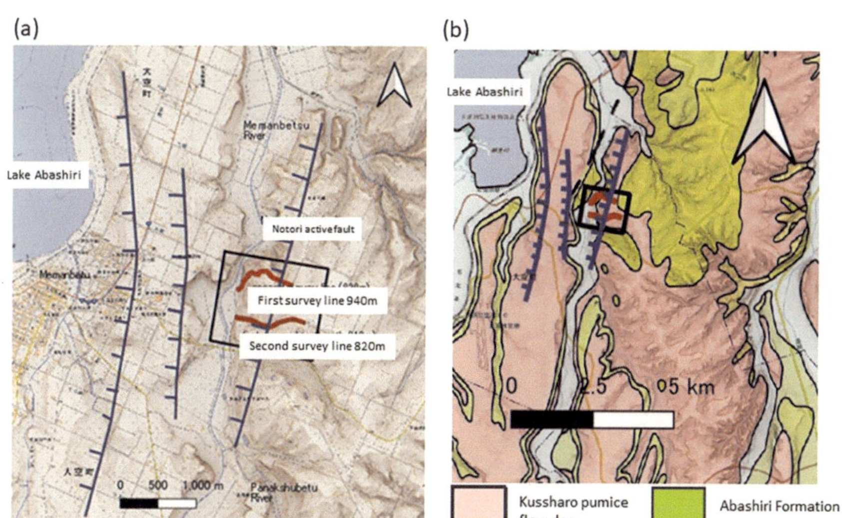

Figure 11.
Topographic map (a) and geological map (b) (edited from 1:200,000 geological map from the geological survey of Japan) of the Memanbetsu area. The dark blue lines show the locations of faults are according to ref. [38]. The black frame indicates the man-borne survey area. The red curve indicates the survey lines.

rim of Lake Kussharo, Japan's largest caldera lake. The altitude of the surrounding area is less than 200 m. The hills are eroded by the Memanbetsu River, which flows north-south and empties into Lake Abashiri (**Figure 11(a)**).

The fault group is part of the Abashiri Tectonic Line, which marks the boundary between the Nemuro and Tokoro belts (**Figure 9**). The Nikoro Group, consisting mainly of basaltic volcaniclastic sedimentary rocks and other formations, belongs to the Cretaceous-Paleogene accretionary complex and is distributed in the Tokoro Belt. On the other hand, the Nemuro Group, composed of Cretaceous-Paleogene sandstone, mudstone, and conglomerate, is distributed in the Nemuro Belt.

These strata are overlain by the Abashiri Formation (Miocene) and the Kusharo Pumice Flow Deposit (Quaternary) (**Figure 11(b)**), among other formations. The Abashiri Formation consists of basaltic andesite and dacite lavas, pyroclastic rocks, conglomerates, sandstones, mudstones, and coeval intrusive rocks. The Kusharo pumice flow deposit consists of dacite pumice and volcanic ash. The study area is covered by the Kusharo pumice flow deposits with an average thickness of more than 10 m. Therefore, the background radionuclide concentrations in GRS are entirely derived from the radionuclides in the Kusharo pumice flow deposits.

3.2.4 Kii-Tanabe area-Honguu active fault-

The survey area is located in the southeastern part of Tanabe City, approximately 110 km southeast of Osaka City (**Figure 9**). The survey area belongs to the Shimanto belt. The Honguu fault is an active fault that runs in an east-northeast direction [38]. The topography on the north side of the fault is an east-west mountain range connecting Mt. Takao (606 m) and Mt. Maki (796 m). The topography on the south side of the fault is a mountainous area oriented north-northeast, including Mt. Kinugasa (235 m) (**Figure 12(a)**). The Right Aizu River and Left Aizu River, which flow north-south and intersect the Honguu Fault, originate in the Hatenashi

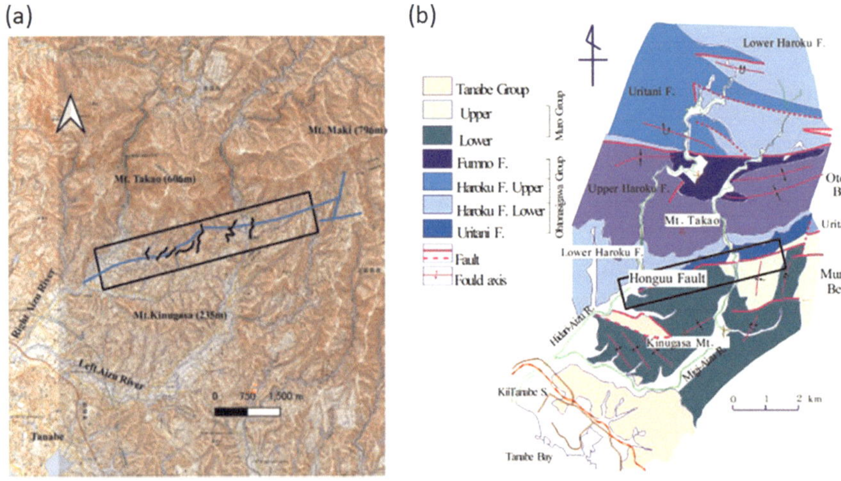

Figure 12.
Topographic (a) and geological (b) maps of the Kii-Tanabe area. The fault of the sky-blue line and geological map (b) are according to ref. [41]. The square frame indicates the survey area. The black curve shows the car-borne survey stop measurement route.

Mountains (the highest point, 1262 m) in the north, join near Kii-Tanabe, and flow into the Pacific Ocean.

The Shimanto belt, which runs from almost east to west, is an accretionary prism formed from the Cretaceous to the middle Tertiary (**Figure 9**). It mainly consists of alternating sandstone and mudstone layers, accompanied by greenstones and chert. The boundary between the Chichibu Belt in the north and the Shimanto Belt in the south is the Butuzou tectonic line. The Shimanto belt is divided into three sub-belts by two tectonic lines. The Gobo-Hagi tectonic line separates the Hidakagawa belt (Cretaceous) in the north from the Otonashigawa and Muro belts (Paleogene) in the south. Furthermore, the Otonashigawa belt and the Muro belt are separated by the Honguu fault (**Figure 12(b)**). The development of the Quaternary system is extremely poor, and only terrace sedimentary layers and alluvium are found in a very narrow area along the river [41].

The Honguu fault divides the study area into northern and southern parts (**Figure 12(b)**). The northern part belongs to the Otonashigawa belt. The southern part belongs to the Muro belt. The fault runs east-west, dips 20–50 degrees to the north, and is accompanied by a fault clay zone several tens of centimeters wide. The Otonashigawa Group of Eocene is distributed in the Otonashigawa belt, and the Muro Group of Oligocene-lower Miocene is distributed in the Muro belt. The Tanabe Group of middle Miocene covers the Muro Group with unconformity (**Figure 12(b)**; [41]). The Otonashigawa Group consists of alternating sandstone and mudstone layers. The Muro Group is composed of sandstone, mudstone, conglomerate, and alternations of sandstone and mudstone. The Tanabe Group consists of an alternation of sandstone and mudstone.

3.2.5 Southern Abukuma area-Tanagura fracture zone and Abukuma belt

The study area is located in the Kuji River lowland area and the southern Abukuma Mountains, covering Hanawa Town and Yamatsuri Town, Fukushima Prefecture. Hanawa Town is about 200 km north-northeast of Tokyo (**Figure 9**). The topography of the study area consists of the Kuji River Lowland and the Southern Abukuma Mountains. The Kuji River lowland in the western part of the survey area belongs to the Ashio (Yamizo) belt, and the eastern part belongs to the Abukuma belt (**Figure 9**). Both belts are separated by the Tanagura fracture zone, which is 2–5 km wide (**Figure 13**).

The Abukuma mountain range is gently undulating, and the elevation gradually decreases from east to west, except for the highlands of the watershed with an altitude of about 890–870 m, which is about 10 km east of the study area. Near the Tanagura east marginal fault of the Tanagura fracture zone, the elevation suddenly drops from 550 m to 300 m (**Figure 13(a)**).

The Ashio belt is underlain by sedimentary rocks of the Paleozoic and granitic rocks. These are unconformably overlain by Tertiary strata. In the southwestern part of the study area, granitic rocks and Tertiary alternating layers of sandstone and mudstone are distributed. The Takenuki metamorphic rocks and the Hanawa plutonic rock mass of the Late Cretaceous are distributed in the Abukuma belt of the survey area. Takenuki metamorphic rocks are high-grade metamorphic rocks of high-temperature type. Hanawa rock mass consists of quartz diorite, tonalite, granodiorite, mylonite granodiorite, and fine-grained diorite [44]. The Hanawa rock mass intruded into metamorphic rocks and sedimentary rocks of unknown age and caused contact metamorphism to them.

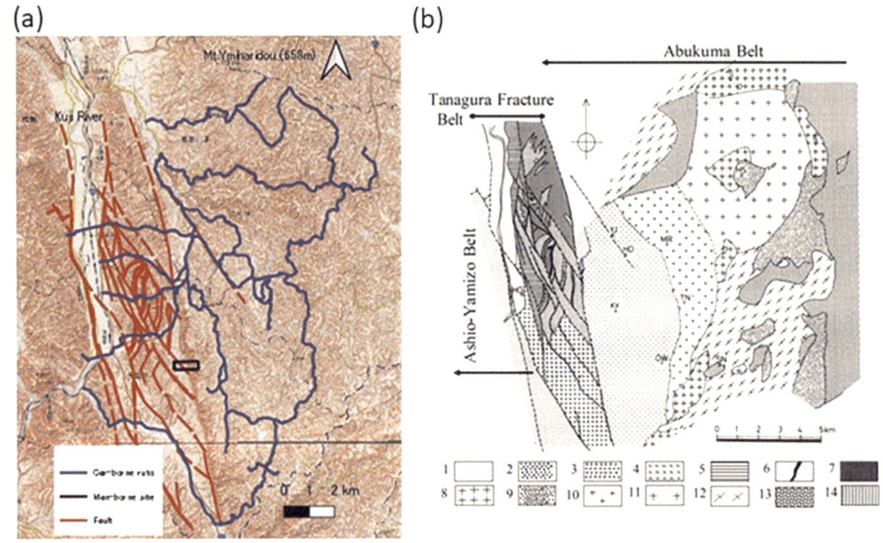

Figure 13.
Topography (a) and geological map (b) of the Southern Abukuma area. The geological map is modified from ref. [42] and Koshiya et al. [43]. Red lines on the topographic map (a) indicate faults from the geological map (b). The black square frame indicates the investigation area of the Tanagura EM fault by man-borne survey. The dark blue curves indicate the investigation route of the car-borne survey. Abbreviations for geological map (b) are as follows: HG: Higashidate, Yujimata, HD: Hidoro, KY: Koya-yachi, MR: Morinoshita, TN: Tonohata, OW: Oiwake, HS: Hashiba, SN: Sankomuroyama. Geological legends for geological map (b) are as follows: 1: Quaternary system, 2: Tertiary system, 3: Higashidate rock body (Mylonitic granodiorite), 4: Onukari rock body (Mylonitic adamellite), 5: Mylonitic and cataclastic sedimentary rocks of the Ashio belt, 6: Cataclastic amphibolite, 7: Mylonitic and cataclastic metamorphic rocks of the Abukuma belt, 8: Fine-grained granular granodiorite, 9: Mylonitic granodiorite, 10: Large sphene-bearing granodiorite, 11: Porphyritic granodiorite, 12: Foliated tonalite and granodiorite, 13: Fine-grained quartz diorite and tonalite, and 14: Metamorphic rocks of the Abukuma belt.

The Tanagura fracture zone is a fault fracture zone bounded by two NNW-trending faults: the Tanagura West Margin Fault and the Tanagura East Marginal Fault (hereafter, Tanagura EM Fault). The following fractured rocks are distributed between these faults: mylonite and fault gouge, which originated from sedimentary rocks, granites, metamorphic rocks, etc. The sedimentary rocks are thought to have originated from the Ashio belt, while the metamorphic rocks are thought to belong to the Takenuki metamorphic rocks of the Abukuma belt [42]. Granitoids are distributed as several small rock bodies. Higashidate and Ohonukari rock bodies are distributed in the study area. The Higashidate rock body is greenish granodiorite. The Ohonukari rock body is gray-white biotite adamellite [42].

Since the Tanagura EM marginal fault can be clearly identified topographically, and the underground geology is known by electrical prospecting [42], it was selected as a target fault for GRS. The geological structure in the vicinity of the survey area has been clarified by many researchers including in Ref. [42] and Koshiya et al. [43]. Here, only the results of photolineament surveys are described. Photolineament interpretation can be seen in various orientations such as NNW, NW, EW, NE, and NNE directions, but the NNW or NW direction is the most prominent (See **Figures 26** and **27** in the Section 4.2). The NNW-oriented lineaments corresponding to the Tanagura EM fault were clearly recognized inside the Tanagura fracture zone area. However, the number of lineaments corresponding to boundary faults of small rock bodies inside the Tanagura fracture zone is smaller than that of the Abukuma zone. The reason for this observation is thought to be that the erosion of the topography in the Tanagura fracture zone made it difficult to interpret the lineament.

3.3 Investigation results

3.3.1 The Atera fault

Man-borne survey, soil radon gas survey, and dipole/dipole array electric prospecting were conducted along the A survey line (extension 660 m), as shown in **Figure 14(a)** [17]. The measurement interval was 20 m. The number of measurement points was 33. According to Yamada [39], survey line A crosses Atera faults A and B. The A fault passes near point 17, and the B fault passes near point 22.

Fresh granite (C_H bedrock) was exposed at survey points 1–2, 7–8, and 30–33 (**Figure 14(a)**). At points 20–26, weathered and sandy C_L bedrock was exposed. Survey point 18 exposed a D bedrock of pale green fault gouge. On the east side of the

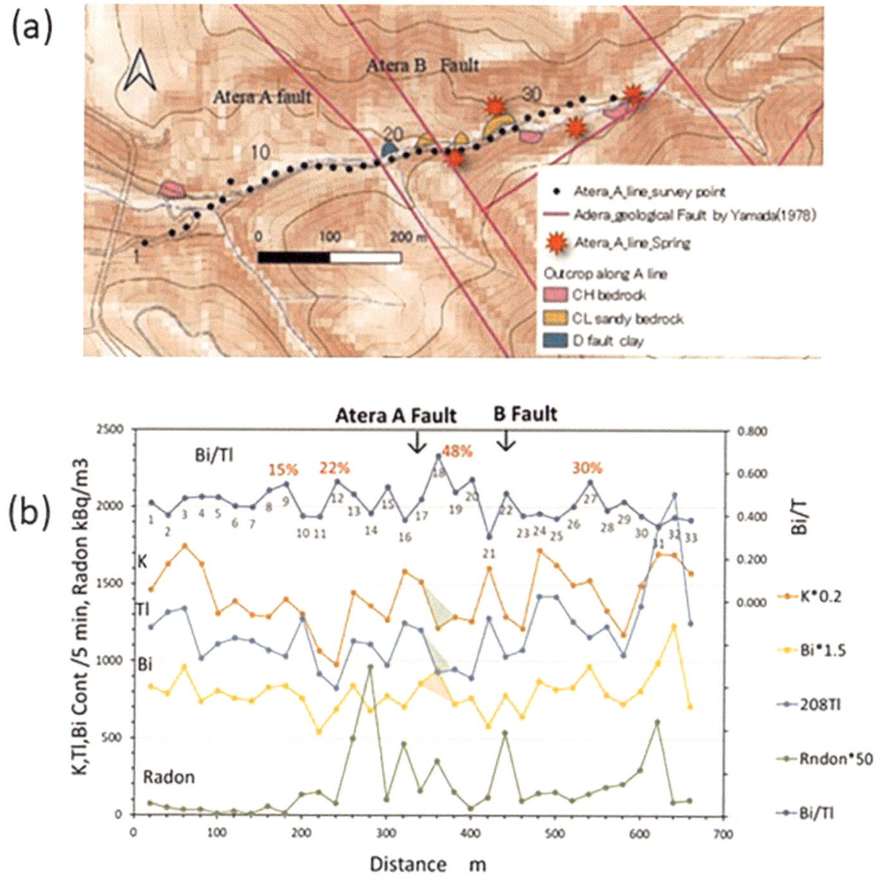

Figure 14.
Location map of survey points and outcrop sketches (a) and fluctuation patterns of K, Bi, Tl, Bi/Tl, and radon concentrations (b) along a survey line across the Atera faults of the Yamaguchi village area. The location of the Atera fault is based on Yamada [39]. The red % number indicates the increasing rate R, and the black number indicates the survey point number. See text for green and red triangles.

Atera B fault, springs of about 1 to 2 liters per minute are distributed in several places (**Figure 14(a)**). Since no springs are distributed on the west side of the fault, it is thought that the fault gouge functions as a water barrier. Based on the above observations, the area between Atera faults A and B is the most fractured.

Figure 14(b) shows the fluctuation patterns of K, Bi, Tl, Bi/Tl, and radon concentrations along the A line. The Atera faults are indicated by arrows in the figure. The largest Bi/Tl peak is at point 18, 20 m east of fault A. The increased rate R of Bi/Tl at point 18 is 48%. This anomaly is composed of the local maxima of Bi (red triangles) and the local minima of Tl (K) (green triangles). Note that there is a fault gouge near point 18.

Since Tl is the least mobile of the three nuclides, the fluctuation pattern of Tl indicates element concentration variation in the original rocks (Section 2.2.4). There should be proportional relationships between Bi, Tl, and K in the original granite. Bi variations deviating from Tl variations were therefore caused by secondary uranium migration and enrichment. In other words, the Bi/Tl ratio allows us to distinguish Bi anomalies into primary magma crystallization anomalies and secondary weathering anomalies occurring near faults. For example, at points 32 and 3, three nuclides, Bi, Tl, and K, are increasing at the same time, so the Bi/Tl variation does not form a peak. Fresh granite distribution is predicted at these locations. In fact, outcrop observations confirmed the distribution of fresh C_H rocks.

Radon concentrations ranged from 0.2 to 19.2 kBq/m^3 with an average value of 3.8 kBq/m^3. The variation pattern of radon concentration along the A line can be divided into two zones. One is the zone where the radon concentration near faults A and B forms several peaks (points 13–31, average 5.4 kBq/m^3). The other is a zone with a low radon concentration and a flat pattern (points 1–12 and 32–33, average 1.2 kBq/m^3) (**Figure 14(b)**).

The highest radon concentration in the former zone is at point 14 (19.2 kBq/m^3), 60 m west of fault A. The third highest peak is at point 22 (10.7 kBq/m^3) near the location of fault B. These peaks are thought to be due to long-range uplifted non-diffusive radon through open fissures associated with faults A and B. However, we need a measured value of a carrier gas such as CO_2 to determine whether it is non-diffusive radon, but there are no carrier gas data. Since the fault gouge is distributed under point 18, the small peak at station 18 is considered to be due to radium near the surface. Point 31 has the second highest peak at 12.2 kBq/m^3. However, since it is a primary anomaly in which three nuclides are high at point 32 at the same time, there is a possibility that this radon is also diffusely transported radon due to radium near the surface.

Granite is distributed on the west side of fault A (points 1–16), and granodiorite porphyry is distributed on the east side of the fault (points 17–33) [39]. Comparing two sides of K (7263:6790), Bi (550: 511), and Tl (1264:1114) shows that the granite distribution area is higher. However, the difference is not large enough to define a fault.

Figure 15(a) shows the variation patterns of Bi/K, Tl/Bi, radon concentrations, and increasing rate of Bi/Tl. **Figure 15(b)** is a simplified pseudo-resistivity profile [12]. The measured resistivity values ranged from 45 to 11,264 Ωm. Since the resistivity of general granite is generally 3×10^2 to 1×10^4 Ωm, various grades of rocks from fault gouge to fresh granite should be distributed on the resistivity profile. The resistivity values are classified into three categories for ease of viewing the resistivity profile in **Figure 15(b)**. Although it is difficult to accurately identify the lithology from the resistivity value, the relationship between the resistivity range and the lithology is as follows, considering the general relationship between the resistivity value and the lithology and field observations: 0–120 Ω·m is from fault gouge to sandy

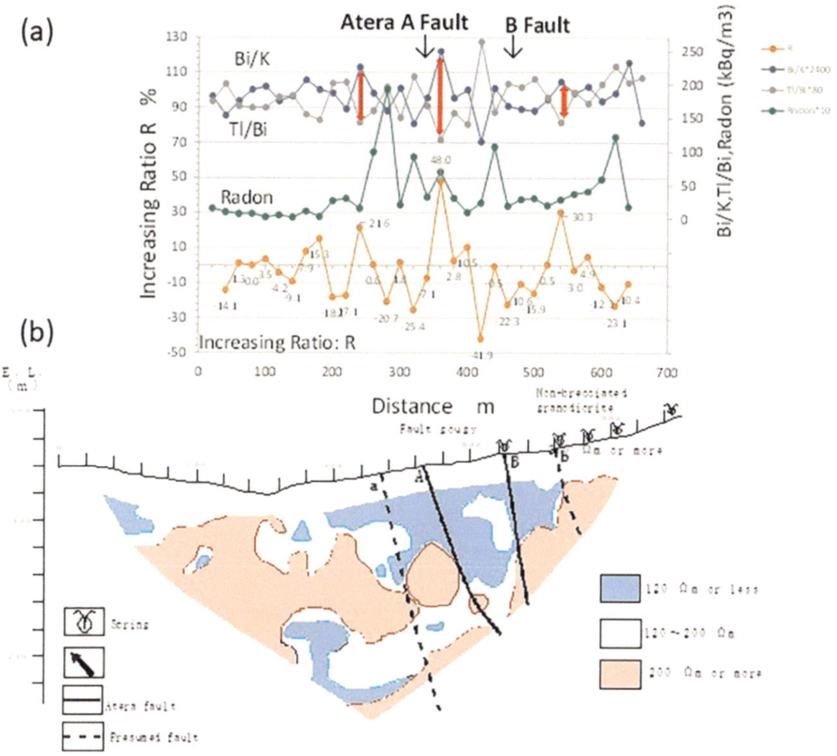

Figure 15.
Fluctuation patterns of Bi/K, Tl/K, radon concentrations and Bi/Tl increasing ratio (a) and the simplified resistivity profile (b) across the Atera faults of the Yamaguchi village area. The numbers on the increasing ratio of Bi/Tl curve indicate the increasing ratio at each point. The resistivity profile has been edited and modified from Imaizumi et al. [12]. The red double-headed arrows in graph (a) indicate the locations where uranium was deposited in the cracks.

granite; 120–200 Ω·m are sandy to brecciated granites; 200–300 Ωm is granite with cracks; and 300–500 Ωm is sheared granite. Above 500 Ωm is fresh granite [12].

Since the Atera fault in the study area is not a boundary fault between two stratums with significantly different resistivity values, it is difficult to draw a fault line in the resistivity profile. The subsurface structure of the Atera fault is inferred from the fault location of geological map and resistivity distribution pattern under the ground surface. Underground traces of fault A and fault B are considered to be shown in the resistivity profile as lines or bands between rock masses of about 300 Ωm or more that have not been broken by fault. Probable fault lines of fault A and fault B are shown in **Figure 15(b)**. According to this rule, unknown faults a and b are also estimated.

The Bi/K and Tl/K curves using SGR logging crack detection methods (Section 2.2.5.3) and resistivity profiles showing subsurface structures are compared in **Figure 15(a)** and **(b)**. The opening positions in both curves (indicated by the red arrows in the figure) show the locations where uranium salts have precipitated in the cracks. At point 18, which has the highest increasing ratio, the opening between the Bi/K and Tl/K curves is the largest (the longest red arrow). Other opening locations (22 and 31% increase) may be related to inferred faults a and b. These faults are "closed fissures" because point 18 is the location of the fault gouge. On the other hand, the radon gas concentration peaks indicating open cracks are located between these closed crack faults.

3.3.2 The Abashiri Lake east coast fault

Man-borne survey, soil gas radon, and CO_2 surveys were carried out along two survey lines, the first line (length 940 m) and second line (820 m), which are set up across the Notori fault (**Figure 11**). The measurement interval was 20 m. Here, fluctuation patterns of K, Bi, Tl, Bi/Tl, radon, and CO_2 along the first survey line are shown (**Figure 16**). For the estimation of fault line (**Figure 17**), the data of the second survey line are also referred. In **Figure 16**, arrows indicate the location of the Notori active fault according to Ref. [38]. The Notori Fault is thought to pass through survey point 25 on the first survey line (**Figure 16(a)**).

The nearest Bi/Tl anomaly to survey point 25 is point 26, 20 m east of the fault, with an increasing rate R of 15%. This anomaly is composed of a small local maximum of Bi (red triangle) and a decreasing trend of Tl. Other anomalies are points 2 ($R = 19\%$) and 14 ($R = 21\%$), which are more than 200 m west of the active fault. These anomalies also consist of small local maxima in Bi (red triangles) and local minima in Tl (green triangles). These two anomalies can be faults because they have the same patterns.

Radon concentrations in soil range from 0.1 to 4.8 kBq/ m^3, with an average value of 1.4 kBq/m^3. Several peaks are formed near station 25 where the active fault passes (**Figure 16(a)**). These radon peaks and CO_2 concentration peaks tend to coincide (**Figure 16(b)**). Therefore, these peaks of radon are non-diffusive radon that has been transported through cracks from deep faults by the carrier gas of CO_2.

The first survey line (as well as the second survey line) has another index indicating the fault. This index is a stepped change in nuclide or nuclide ratio that indicates different material layer boundaries (stratum boundaries). The fluctuation patterns of Bi, K, and Bi/Tl show a sharp decrease at points 29–30 (indicated by green dotted line). Averages of K, Bi, and Bi/Tl at points 0–29 and at points 30–47 are as follows: 3563:2482 on K, 315:235 on Bi, and 0.56:0.46 on Bi/Tl (**Figure 16(a)**). Therefore, the fault shown by the stepped change index passes through point 30. Although the survey site is covered with the Kussharo pumice layer with an average thickness of 10 m, the stepped change in Bi/Tl may indicate the boundary fault of the different

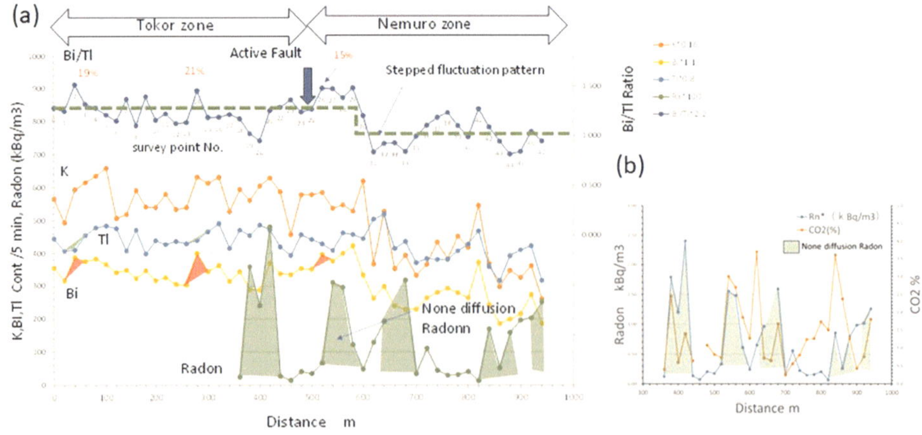

Figure 16.
Fluctuations of K, Bi, Tl, Bi/Tl, and radon concentrations (a) and fluctuations of CO_2 and radon concentrations along the first survey line across the Notori fault of the Memanbetsu area.

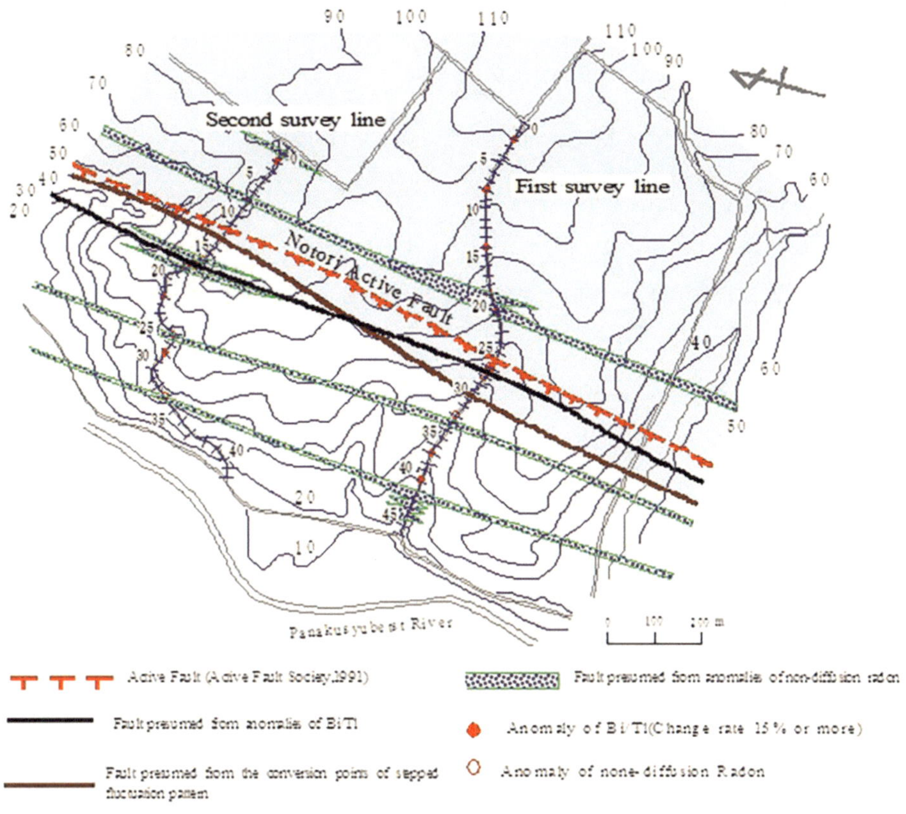

Figure 17.
Relationship between presumed faults and the active faults of the Memanbetsu area.

Kussharo pumice layers, which have different nuclide concentrations. Since the Kussharo pumice layer sandwiches a thin sand layer, the different nuclide concentrations may be due to the different depths of the sand layer.

Figure 17 shows the relationship between the inferred faults and active faults along the first and second survey lines. The two inferred faults are the Bi/Tl anomaly points linking fault (black line) and the Bi/Tl stepped fluctuation of turning points linking fault (brown line). Both the faults estimated by the two methods are shifted 20–100 m to the west from the active fault. The non-diffusive radon peak linking faults are distributed as several crack zones 100–250 m apart to the east and west of the fault.

3.3.3 The Honguu fault

A two-stage survey, reconnaissance and detailed survey, was conducted to locate the Honguu fault. In order to extract seven routes for detailed survey, the running car measurement of car-borne was carried out along 19 routes around the Honguu fault and a total survey distance of 64.8 km. For the detailed survey, the stopping car measurement of car-borne was performed (**Figure 18**). The total distance surveyed was 4930 m. This paper describes the results of the C5 line (**Figure 19**) that localized the Honguu fault using GRS and soil radon gas surveys.

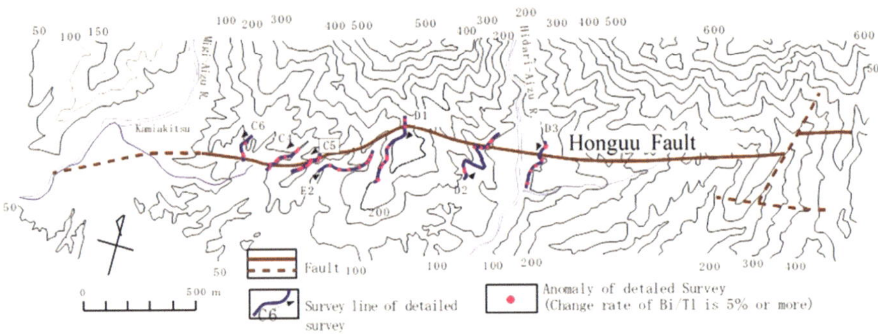

Figure 18.
Detailed car-borne survey route and distribution of Bi/Tl anomalies of the Kii-Tanabe area.

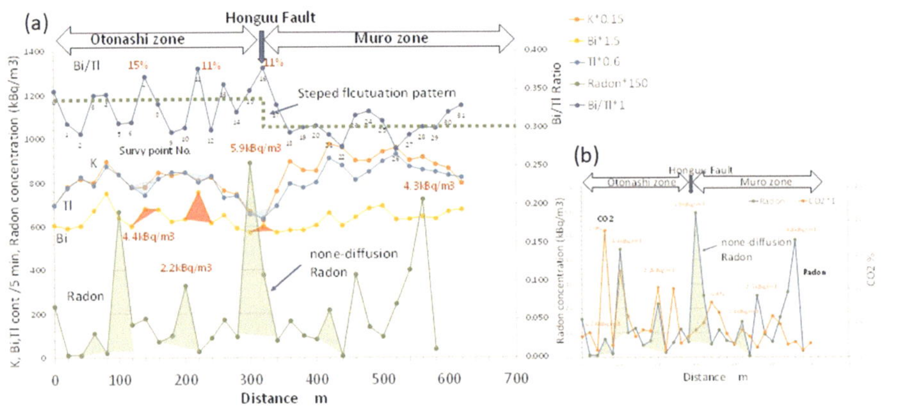

Figure 19.
Fluctuations of K, Bi, Tl, Bi/Tl, and radon concentrations in soil (a) and fluctuations of radon and CO_2 concentrations in soil along the C5 survey line across the Honguu faults of the Kii-Tanabe area. See the text for an explanation of hatching patterns and green dotted line.

Figure 19(a) shows fluctuation patterns of K, Bi, Tl, Bi/Tl, and soil radon concentrations. According to the ref. [41] and topographical interpretation of aerial photographs, the Honguu fault is considered to pass through survey point 16. Point 16 has a Bi/Tl anomaly (increasing rate R = 11%). There are two other anomalous points, 100 m north of the fault (point 11, R = 11%) and 180 m north (point 7, R = 15%). All anomalous points are in the Otonashi Belt and not in the Muro Belt. The cause of this observation is unknown. These anomalies are all formed from Bi local minima (red triangles) and Tl local minima (blue triangles), which should indicate fault patterns.

Radon gas concentration ranges from 0.05 to 11.1 kBq/ m^3 with an average of 1.7 kBq/ m^3. There are several peaks around the Honguu fault. The non-diffusive radon peaks that overlap with the soil CO_2 peak (**Figure 19(b)**) have the highest concentration peak at point 15 (5.9 kBq/ m^3), 20 m north of the fault. Other peaks include point 10 (2.2 kBq/ m^3), 120 m north of the fault and point 5 (4.4 kBq/m^3), 220 m north of the fault. These peaks are not related to the Honguu fault, but are related to the derived faults estimated from the Bi/Tl anomaly.

The Honguu fault is also the boundary fault between the Otonashigawa group and the Muro group, which have different sediment ages and nuclide concentrations. In fact, the mean values of K, Tl, and Bi/Tl at survey points 0–16 and 17–31 differed: 5229:5927 on K, 1295:1395 on Tl, and 0.33:0.30 on Bi/Tl, respectively. The stepped change in Bi is indicated by the green dot line.

Figure 18 shows the distribution of anomalies with a Bi/Tl increase rate of 5% or more. Due to the small threshold for fault detection on all survey lines, the anomaly points are distributed not only near the Honguu fault but also at locations far from the fault. Strict threshold settings are required to eliminate these extra (noise) anomalies. This issue is considered in Section 5.

3.3.4 The Tanagura east marginal fault

The Tohoku Agricultural Administration Office (hereafter, TAAO) evaluated bedrock groundwater development technologies, such as fault analysis, electrical surveys, and groundwater geochemical surveys, in rocky areas where groundwater development is difficult [42]. The southern Abukuma area, where various rock bodies are distributed, was selected as the test site. The author was in charge of the fault crack survey using radioactive prospecting. This section presents reviewed results of man-borne survey and soil radon gas surveys along the survey line across the Abukuma EM fault. The results of a regional running car-borne survey are described in the next section. TAAO [42] conducted an electrical survey of 900m survey line across the Tanagura EM Fault

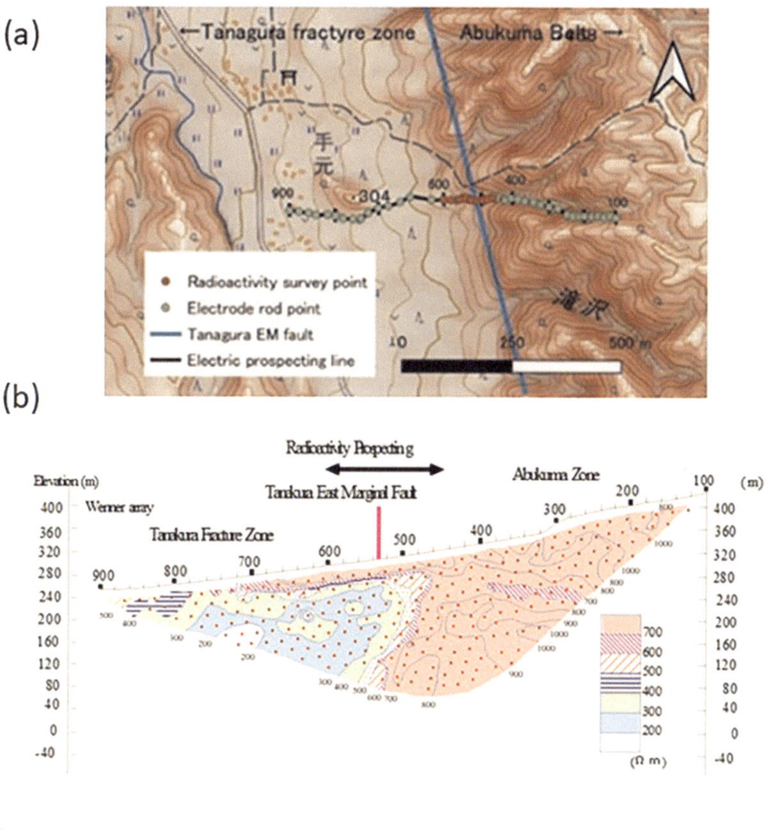

Figure 20.
Electric prospecting survey line and radioactivity prospecting survey point (a) and apparent resistivity profiles by the Wenner electrode array across the Tanagura east marginal fault (b) of the Southern Abukuma area. The resistivity profile from the TAAO [42] has been revised and added.

(**Figure 20**). The resistivity profile shows that the fault gouge and mylonite of granodiorite origin (resistivity 400 Ωm or less) are distributed on the west side of the fault (Tanagura fracture zone) and the mylonitized granodiorite (700 Ωm or more) is distributed on the east side of the fault (Abukuma zone). TAAO [42] concluded that the fault underlies under the survey line 530 m. Gamma ray spectrometry and radon gas survey were carried out along the electrical survey line of 450–600 m. The measurement interval was 10 m.

The fluctuation patterns of K, Bi, Tl, Bi/Tl, and radon concentrations along the survey line are shown in **Figure 21**. These patterns can correspond to the fluctuation patterns of the F dam site (**Figure 7**) where U leaching occurred in the fault fracture zone. Bi begins to decrease from both sides of the fracture zone (survey lines 500 and 590 m) and forms a minimum in the center part of the fracture zone. Comparing the variation patterns of K, Bi, and Tl, only Bi decreased (gray triangles). The minimum occurred at 550 m. This location is the center of the fracture zone indicated by GRS. The local maximum value of Bi (red triangle) caused by U enrichment is at 500 m at the entrance of the fracture zone, where Tl is a local minimum (blue triangle). Remember that the typical fracture zone shows fault clay layers form at both ends of the fracture zone (Section 2.2.5.1). U enrichment might have caused U to get adsorbed in these clay layers. As a result, Bi/Tl formed the only anomalous point (orange triangle) with an increasing rate of 11% along the survey line.

Radon concentrations range from 0.4 to 2.5 kBq/ m^3 with an average value of 0.9 kBq/ m^3. The fluctuation pattern of radon concentration is also the same as that of F dam. There is a single peak of 2.5 kBq/m^3 at line 520 m, 10 m east of the fault location. It is not possible to determine whether this peak is non-diffusive radon

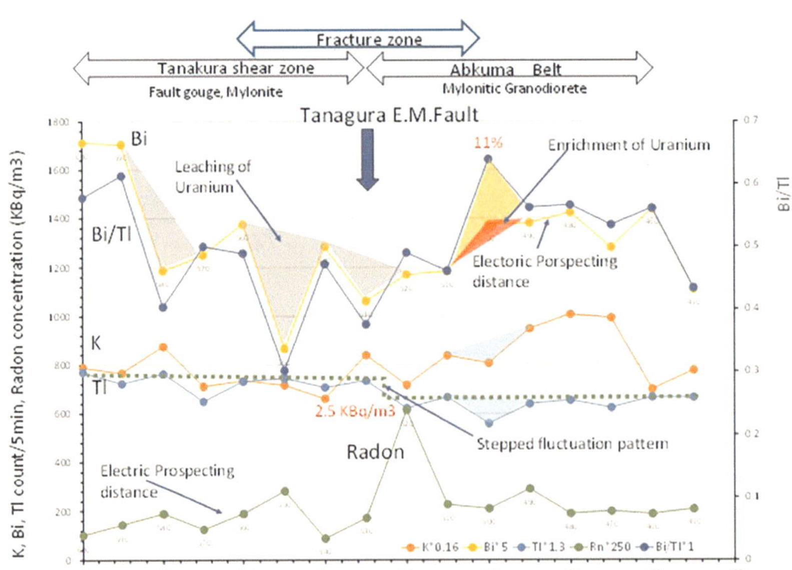

Figure 21.
Fluctuation patterns of K, Bi, Tl, Bi/Tl, and radon concentrations in soil along the electrical prospecting lines 450 to 600 m across the Tanagura east marginal fault of the southern Abukuma area. The position of the electric survey line is the number attached to the curve of Bi and radon. Red numbers are the increasing rate of Bi/Tl and radon concentration at the peak point, respectively. The Tanagura EM fault passes through the 530 m electric prospecting line.

DOI: http://dx.doi.org/10.5772/intechopen.1002936

because no CO_2 measurements were made in this study, but the only peak is thought to be radon moving upward through an open crack associated with the fault.

The Tanagura EM Fault is a boundary fault between two geological zones with different ages and nuclide concentrations. Therefore, there should be a difference in nuclide concentration on both sides of the fault. In fact, Tl count increases stepwise around line 530 m. **Figure 21** shows this change with the green dotted line. The Tl averages of 450–520 m and 530–600 m are 490 and 1560, respectively. Tl count increases threefold west of this fault. Since Tl is the most resistant element to weathering, this stepped change is considered to indicate that fault movements caused rocks of different compositions to come into contact.

4. Regional crack distribution survey by car-borne survey

4.1 Atera fault car-borne survey

In order to investigate the regional distribution of the Atera faults, the running car measurement of car-borne was carried out on 19 routes across the Atera fault [17]. Total survey distance is 50.1 km (**Figure 10**). Car-borne anomalies were defined by the Bi/Tl increasing rate R above the threshold. Car-borne survey recorded the integrated value for 30 s at 4 km/h running, so the measurement interval is 30–40 m. Anomaly locations were plotted at the center of the measurement interval. The reproducibility of the position of the anomaly was confirmed by the stopping car measurements at intervals of 10 m around several anomalies.

To determine the R threshold, we plotted anomalies with R = > 20, 20–15%, and 15–10% on a geological map and examined the relationship between the anomaly distribution and known fault locations (**Figure 22**). The number of Bi/Tl anomalies

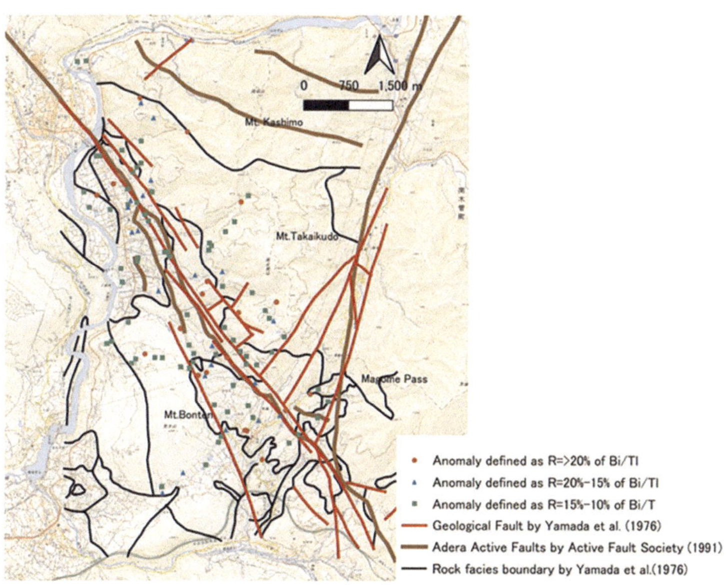

Figure 22.
Relationship between the distribution of anomalies and the locations of the geological faults and active fault of the Yamaguchi village area.

corresponding to faults is 4 for R= > 20%, 5 for R = 20–15%, and 16 for R = 15–10%. Of the R= > 20% anomalies near Magome Pass, only one corresponds to the Atera fault. If the threshold is set to 10% or more, the correlation between faults and anomalies will improve, but noise anomalies will increase. Since multiple anomalies are required for fault estimation, the Bi/Tl threshold for the study area was set at R > =10%.

The anomaly point is calculated from the ratio of the average value (called Bi/Tl background here) of five intervals (150–200 m) before and after the measured value. It is presumed that the wide-area distribution of the background also has fault information. **Figure 23** shows the relationship between the Bi/Tl background, anomaly points, faults, and topography. The background ranged from 0.46 to 0.20. No association was observed between background values and the distribution of anomalies. Four high background areas with Bi/Tl > 3.6 are distributed: Northwestern part (1) and central part (2) of the survey area, northwestern slope of Mt. Takadoki (3), and near Magome Pass (4) (**Figure 23**).

Area (1) is distributed in an elongated area parallel to the Atera fault. Area (2) has two peak points of Bi/Tl = 3.9. A subarea with one peak is concordant with the Atera fault. Another subarea with another peak is concordant with a secondary fault from the Atera main fault. The area near Magome Pass is distributed on the Magome-Pass fault that intersects the Atera fault. These regions are located in terrain erosion zones caused by faults and are thought to be related to faults (**Figure 23(b)**). While region (3) is distributed on high slopes away from faults, the formation mechanism of this region (3) is unknown. On the other hand, the low background area with Bi/Tl > 2.6 extends from the southern slope of Mt. Takadoki to the Atera fault lines. The relationship between this low Bi/Tl area and the fault is unknown. According to the weathering migration mechanism of U, the Bi/Tl background may also be related to U leaching and enrichment. The high background areas of (1) and (2) may be related to U enrichment along the Atera fault fracture zone.

Figure 24 shows the fluctuation patterns of the nuclides and nuclide ratios along the survey line in the vicinity of Magome [17]. The arrow shows the fault location from the geological map. The anomalies of Bi/Tl (R = 32%) and Bi/K (R = 26%) were

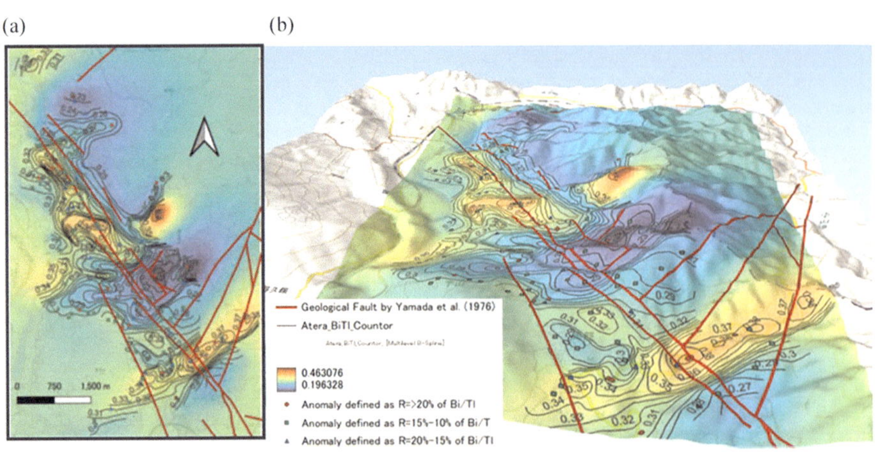

Figure 23.
Relationship maps between the Bi/Tl background, anomaly points, faults and topography of the Yamaguchi village area: (a) Plan view, (b) 3D view. The Qgis2threejs plugin for three-dimensional (3D) display of quantum geographic information system (QGIS) projects the fault and Bi/Tl background distribution on the 3D terrain.

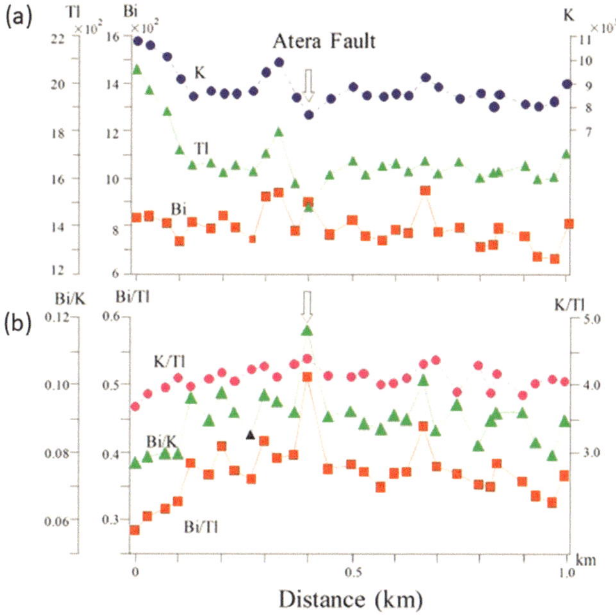

Figure 24.
Fluctuation patterns of the nuclides and nuclide ratios on the survey line around the Magome of the Yamaguchi village area (modified from Imaizumi et al. [17]). The arrow shows the fault location from the geological map [39].

formed by the local maximum of Bi and the local minima of Tl and K. Most of anomalies by the car-borne survey related to the known faults showed the same fluctuation patterns. It is worth noting that the fault-indicating nuclides and nuclide ratio patterns observed by man-borne survey also hold true for the 30 m of car-borne survey interval.

The radioactive anomalies of Type I might be formed by contact metamorphism near the boundaries of granite bodies (**Figure 2**). Possible zones of Type I are the boundary between granodiorite and granite, and the boundary between rhyolite and granodiorite. These zones are indicated by dotted lines in **Figure 10(b)**. In fact, there is an anomaly of R > = 20% on the lithofacies boundary at the southern foot of Mt. Sizumo in the northern part of the study area. In the southern part of Mt. Takadoki and the western part of Mt. Bonten, there are anomalies with R > =10% on the lithological boundaries (**Figure 10(b)**). These anomalies are thought to be caused by Type I. **Figure 25** shows the fluctuation patterns of the nuclides and nuclide ratios on the survey line across the boundary of rock facies between granodiorite porphyry and granite around Mt. Sizumo [17]. The Bi/Tl (R = 41%) anomaly is formed only by an increase in Bi. This pattern differs from that of the fault.

Figure 22 shows many anomaly points that cannot be explained by known faults or contact metamorphic boundaries. Some of these anomalies may represent hitherto unknown faults if they have the patterns of faults. Other anomaly points may be noise.

4.2 South Abukuma car-borne survey

Groundwater is expected to flow through the fracture system. In Tohoku Agricultural Administration Office project [42], the purpose of the car-borne survey was to

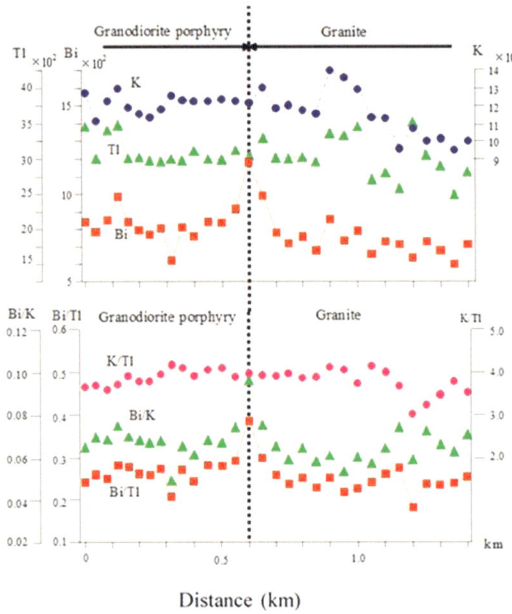

Figure 25.
Fluctuation patterns of the nuclides and nuclide ratios on the survey line across the boundary of rock facies between granodiorite porphyry and granite around Mt. Sizumo of the Yamaguchi village area (modified from Imaizumi et al. [17]).

verify its ability to detect fracture systems. In order to achieve the purpose, 30 routes and a total distance of 121 km of car-borne survey were carried out. The anomaly points of Bi/Tl as fracture signs were defined as R > =15%. The possibility of detecting faults by car-borne survey was evaluated based on the degree of agreement between the distribution of car-borne anomalies and the photolineaments [42]. In addition, the geological classification ability of car-borne survey was evaluated from the relationship between lithofacies boundaries and anomalous points and clay mineral estimation by K/Th plots.

4.2.1 Degree of agreement between the distribution of car-borne survey anomalies and the lineaments

Figure 26(a) shows the relationship between anomalous points and lineaments. The total number of anomaly points is 79. Forty-nine percent of all anomaly points are associated with lineaments. The rate of coincidence with lineaments is low (13%) within the Tanagura fracture zone because there were few lineaments within the Tanagura fracture zone (see Section 3.2.5). Considering only the Abukuma belt, 58% of the anomalies are associated with lineaments. Anomalies within the Tanagura fracture zone tend to coincide with boundaries of rock masses delimited by faults (See next section).

Figure 27(a) shows the relationship between photolineaments (771 lines) [43] and lineaments crossing the car-borne survey routes (203 routes). The predominant directions of the photolineaments and the car-borne survey lineaments show almost the same direction. This figure shows that the car-borne survey was uniformly surveyed over the area.

The directionality of the lineaments that coincide with the anomalous points was investigated by the following statistical method. This method was to classify the

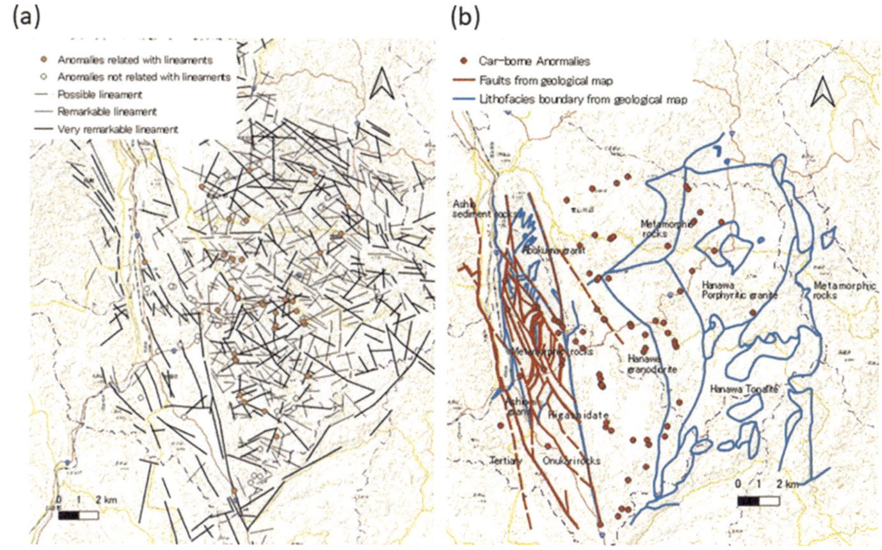

Figure 26.
Relationship between photolineament [42, 43] and car-borne survey anomalies (a) and relationship between the geological boundary [42] and car-borne survey anomalies (b) in the Southern Abukuma area.

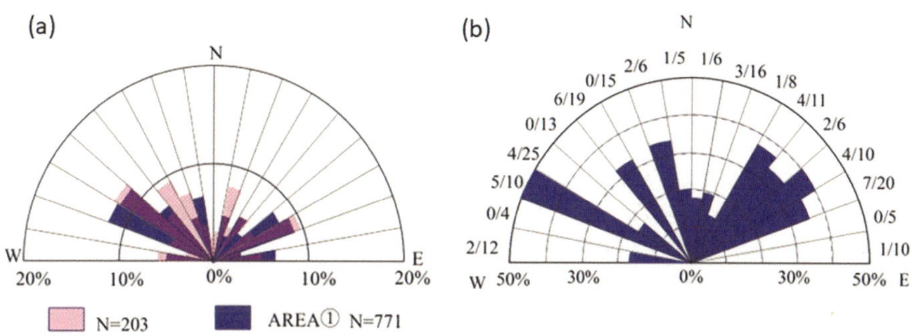

Figure 27.
Frequency diagrams of the direction of photolineaments (modified from Koshiya et al. [43]) and photolineaments intersecting the car-borne survey lines (a) and frequency diagram of lineaments related to anomaly points (b) in the Southern Abukuma area. The frequency diagram of photolineaments shows the diagram of area 1 by Koshiya et al. [43]).

direction of the lineament every 10° and examine the number of cases that coincide with anomalous points. The concordance rate of lineaments every 10° that coincided with the anomalous point was investigated using the following equation.

$$\textit{Concordance rate} = \frac{\textit{Number of coincidences between lineaments and anomaly points}}{\textit{Number of intersections between lineaments and carborne routes}} \times 100 \tag{3}$$

Figure 27(b) shows the directional frequency distribution of lineaments with a high matching rate. The direction of lineament with high matching rate is N60°-50°W (50%), N30°-70°E direction (35%), N10°-20°W direction (33%), and N30°-40°W direction (32%). This trend is the same as the frequency distribution map of lineaments in **Figure 27(a)** [42].

4.2.2 Relationship between car-borne survey anomaly points and rock bodies

Many anomaly points within the Tanagura fracture zone tend to be plotted on the fault line (**Figure 27(b)**). On the other hand, in the Abukuma belt, unlike the Atera fault area (Section 4.1), with some exceptions, the anomalous points do not coincide with the lithofacies boundary. They are scattered in each rock body (**Figure 27(b)**). Since uranium enrichment in the contact metamorphic zone does not occur through the intrusion of granite into the high-temperature host rock deep underground (**Figure 2**), the granites of the Abukuma Belt may have intruded at a deeper depth than those of the Atera area.

4.2.3 Evaluation of rock weathering by K-Tl cross-plot

In order to plot GRS data on the K-Th cross-plot, a calibration pad as defined by the IAEA [4] is required. Unfortunately, due to the lack of calibration pads available in Japan, it is not possible to convert γ-rays K and Tl to potassium oxide (K2O) and Th concentrations by formal methods. Therefore, γ-rays K and Tl were converted into concentrations by the following simple method. Japan's National Institute of Advanced Industrial Science and Technology publishes geochemical maps of all of Japan on the Internet (https://gbank.gsj.jp/geochemmap/gmaps/map.htm). According to this information, the study area is within the areas of K2O = 1.609–1.887% and Th = 11.60–14.74 ppm. Assuming that K2O = 1.609%, Th = 11.60 ppm, and the mean values of γ-rays K and Tl are linearly proportional to these values, γ-rays K and Tl were converted to each concentration datum.

Figure 28 shows the correlation of K_2O-Th for each rock body in the Tanagura fracture zone, except for the Hanawa rock body which belongs to the Abukuma belt. **Table 2** shows regression equations and clay mineral species for each rock body using **Table 1**. According to the Th-/K cross-plot (**Figure 3**), clay minerals in each rock mass are plotted between the kaolinite line (K/Tl = 12) and the mixed layer clay line (K/Tl = 3.5). Therefore, the main clay minerals in all rock bodies are mixed layer clays. However, parts of the Neogene, Ashio sedimentary rocks, Metamorphic rocks, and Ohonukari rock bodies are plotted in the montmorillonite region. From **Table 1**,

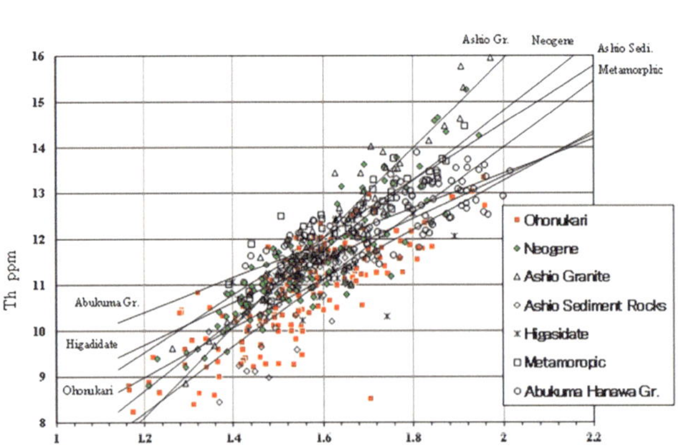

Figure 28.
The correlation of K_2O-Th between Neogene sedimentary rocks, Ashio granites, Ashio sedimentary rocks, Ohonukari rocks, Higashidate rocks, metamorphic rocks, and Abukuma Hanawa granites in the Southern Abukuma area.

Geologic province	Rock facies	Regression equation	R^2	Clay mineral
Tanagura zone	Ashio granite	y = 9.7654x-3.5928	0.93	Kaolinite and chlorite
	Neogene Sediment	y = 7.630 Sx- 0.4579	0.84	Mixed-layer clays: Montmorillonite
	Ashio sediment Rocks	y = 7.2368x-0.4683	0.53	Mixed-layer clays: Montmorillonite
	Metamorphic Rocks	y = 6252 5x + 2.0278	0.81	Mixed-layer clays: Montmorillonite
	Ohonurai Rock Mass	y = 5353 5x + 2.5684	0.60	Mixed-layer clays: Montmorillonite
	Higashidate Rock Mass	y = 4.587 7x + 4.1897	0.41	Mixed-layer clays
Abukuma Belt	Abukuma Hanawa Granite	y = 3.7866x + 5.8611	0.58	Mixed-layer clays: Illite

Table 2.
Regression equations of K_2O-Th and clay mineral species estimated for each rock body in the Southern Abukuma area.

kaolinite and chlorite may be found in the Ashio belt origin granite (K/Tl = 9.8), and illite may be found in the Abukuma belt granite (K/Tl = 3.8). Summarizing the results from the figure and table, the most weathered rock body is the Ashio belt origin granite distributed in the Tanagura fracture zone. The least weathered rock is the Hanawa rock body of the Abukuma belt.

5. Discussion

5.1 Characteristics of distribution patterns of nuclides around faults and fault detection indices

Due to the ease of exploration, the anomaly defined by the Bi/Tl increased rate and the radon gas concentration peak were selected as indicators for fault trace detection. We evaluated whether these indices can detect four existing buried faults. The results are summarized in **Table 3**. Figures in the table show the distance from the locations of the already-known fault. The three kinds of faults presumed by radioactivity prospecting are: (1) Fault presumed from the continuity of anomaly points with an increasing rate Bi/Tl above the threshold defined for each study area. (2) Fault estimated from the continuity of peak points of non-diffusive radon (more precisely, the radon has to coincide with CO_2 gas peak points). (3) Faults inferred from the continuity of conversion points where the average value changes in the stepped fluctuation pattern of nuclide and nuclide ratio. Point (3) was newly added based on field survey results.

The index of (1) was near all four faults. In particular, it should be emphasized that the location of the Bi/Tl index was within 0–30 m of the known fault location. Anomalies of Bi/Tl are not due to an increase in Bi alone. They are caused by relative changes in Bi and Tl. All fault-related anomalies were formed by local maxima in Bi and local minima or decreasing trends of Tl. The geological meaning of the index (1) is considered to be related to a fault gouge, although only the Adela fault has been confirmed. The mechanism (1) is explained as follows: Uranium salts in groundwater were adsorbed on clay minerals (e.g., montmorillonite: see Sections 2.2.4 and 2.2.5) in the fault gouge. On the other hand, the concentration of Th in the fault gouge tends to preserve the composition of original rock (see Section 2.2.4). Given the above mechanism, we can understand the limitations of

	Fault name	Geology	Survey line	Fluctuation pattern ^{214}Bi	^{208}T1	^{40}K	Index (1) ^{214}Bi/^{208}Tl	Geology of Anomaly	Index (2) Non-diffusive ^{222}Rn	Index (3) Stepped fluctuation
Active Fault	Atera Fault	Granodiorite / Granite	A	∧	∨	∨	∧ 20 m	Fault gouge	○ 60 m	×
	East Abashin Lake Faults	Tertiary volcanic rocks / Tertiary volcanic rocks	First	∧	↘	↘	∧ 20 m		○ 180 m	○ 100 m
			Second	∧	↘	↘	∧ 20 m		○ 120 m	○ 0 m
	Honguu Fault	Tertiary Otonashigawa F.shale / Tertiary Muro F. alternation	C5	∧	∨	∨	∧ 0 m		○ 10 m	○ 0 m
Non-active fault	Tanakura East Marginal Fault	Granitic rocks/ Mylonite, Fault gouge	Electric prospecting	∧	∨	∨	∧ 30 m		○ 30 m	○ 0 m

∧*: Maximum,* ∨*: Minimum,:* ↘ *Decrease. Figures show the distance from the location of faults.*

Table 3.
Summary table of fault detection by radioactivity prospecting.

radioactivity prospecting. Bi/Tl anomalies cannot detect existing fault locations, such as when fault gouging is insufficient in fault fracture zones.

Index (2) was confirmed for all faults. The locations of the (2) index were within 10–180 m of the known fault location. In this paper, the CO_2 survey results are shown only for two faults, but radon concentration peaks that coincide with the CO_2 peaks were shown for these two faults. These prove the existence of non-diffusive radon. In a typical fracture zone structure, the open cracks develop around the central fault gouge (**Figures 6** and **7**). The peak position of (2) may indicate an open crack (open fault). Therefore, the peak position in the fracture zone does not match the Bi/Tl anomaly position located in the fault gouge. In the Notori fault of Abashiri Lake East Fault Group, several lines of the non-diffusive radon peak linking faults were distributed 100–250 m apart to the east and west of the fault (**Figure 17**).

The fault estimated by the index (3) was confirmed at three locations, except for the Atera fault. This phenomenon occurs when strata with different nuclide concentrations are in contact with each other at the fault boundary, so it is thought that it was not detected at the boundary between granodiorite and granite on the Atera fault. The positions of the turning points of the three faults were observed within 0–100 m from the known faults.

Since the Honguu fault and the Tanagura EM fault contact different rock bodies (stratums) with their faults, the fault detection by the index (3) is expected. However, for the Abashiri Lake Toho Fault (Notori Fault), the fault could be detected with this index, even though the Kusharo pumice flow deposits are distributed on both sides of the fault. This phenomenon was explained by the difference in the depth of thin layers (for example, sand layers) sandwiched between pumice flow deposits, but it is necessary to clarify the cause in the future.

The radioactivity fluctuation pattern of the fault at the F dam site, which has a fracture zone width of about 20 m, is considered to be the simplest fluctuation pattern. However, it is worth noting that fracture zones with a width of 50 m or more, such as the Tanagura electromagnetic fault, also have a similar deformation pattern to the fault at the F dam site (**Figure 21**). This phenomenon is the basis for detecting faults with the Bi/Tl index, even in running carbon with wide measurement intervals (**Figure 24**). However, as shown on the Notori fault (**Figure 17**), large shatter zones cannot be detected by single gamma-ray anomalies or radon anomalies alone. It should be kept in mind that in general, large fracture zones have a complex distribution of closed and open cracks.

In summary, it is possible to detect fault traces with three indices. However, each index has an error of 0–30 m, 0–180 m, and 0–100 m from the published fault position, respectively. In particular, index (2) may exist in multiple locations within the fracture zone, and even outside the fracture zone (Section 2.2.5.1).

5.2 Relationship between car-borne survey anomalies, faults, and lithofacies boundaries

Car-borne survey is an exploration technology that bridges the gap between man-borne and air-borne [4]. Even if car-borne survey is carried out in off-road areas, geological mapping using car-borne survey may not be able to investigate the original geological situation because the roads are heavily contaminated with materials used in road construction [4]. Therefore, the main purpose of a car-borne survey is recommended to be applied to environmental problems such as searching for lost radioactive sources and mapping radioactive fallout [4]. However, Kimura's car-borne system (Section 3.1.2) was able to extend the limits of car-borne survey by giving direction to the measurement efficiency of the detector (Fault/lineament detection: Section 4.1, 4.2.1, geological boundary mapping: Section 4.2.2, weathering assessment: Section 4.2.3). Directivity to the measuring efficiency of the detector is given by installing the detector on one side of the cargo hold and shielding one side of the detector.

The distribution of abnormal points, which are a fault signature, changes depending on the threshold set arbitrarily by humans. Therefore, if the threshold value is set too small, noise anomalies other than faults and geological information will occur. Therefore, noise processing methods are one of the future challenges. Artificial Intelligence (AI) machine learning can be expected to become one of the processing methods.

Machine learning models such as random forests can draw decision boundaries for linearly inseparable distributed anomalous and normal points. Using random forest classification (OpenCV), which is a built-in function of SAGA-GIS, an attempt will be made to classify the car-borne survey anomalies in the Atera fault zone.

We already have location information of anomaly point distribution with R > =10% in GIS. In order to draw abnormal point areas in GIS, location information of normal areas without faults is required. For simplicity, the distribution of normal points without faults was created using the QGIS plugin "QChainage." The QChainage can create points at specified intervals on the lines of the car-borne survey route on QGIS. Here, points were generated every 200 m on the car-borne survey routes (254 points). Among these points, points that overlapped with the anomaly points and points within 200 m of the anomaly points were deleted to obtain the location information of normal points (167 points) (**Figure 29**).

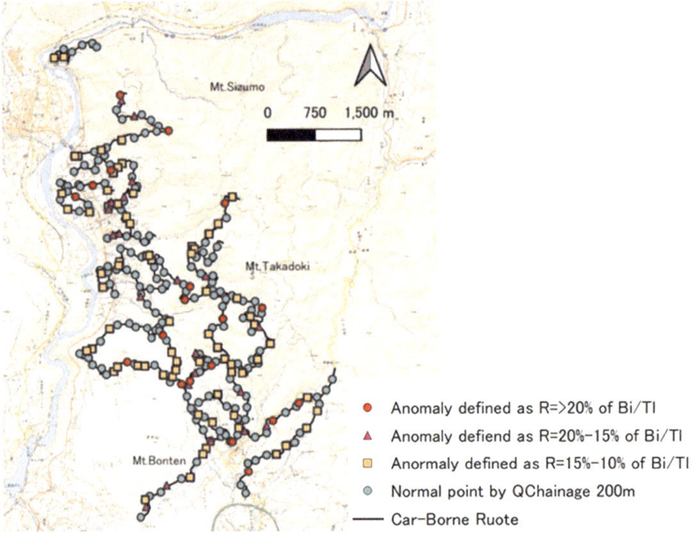

Figure 29.
Distribution map of anomaly and normal points in the Yamaguchi village area. The anomaly points were defined as R > =10%. The distribution of normal points was created using the QGIS plugin "QChainage." see text for detailed explanation.

Training data for Random Forest classification (OpenCV) require raster data. The point data of abnormal points and normal points were converted to raster data using the IDW function (Inverse Distance Weighted). **Figure 30** shows the classification results of anomaly and normal areas. The anomaly points area is indicated by a black polygon. In this classification, only the location information of abnormal points and

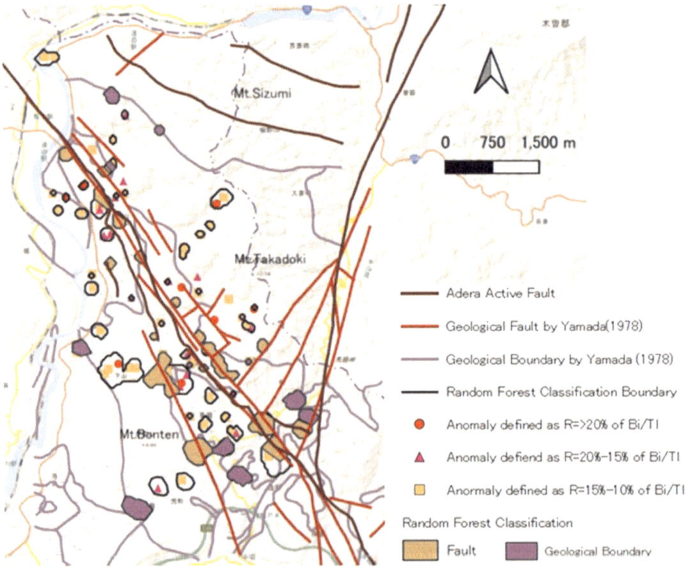

Figure 30.
Results of classification of anomaly and normal regions using the SAGA-GIS built-in function random Forest (CV) in the Yamaguchi village area. Anomaly areas related to faults (brown areas) and anomaly areas related to geological boundaries (purple areas) are manually colored.

normal points was used as a covariate, so only the areas of abnormal and normal points could be classified. Therefore, in this figure, anomaly areas related to faults (brown areas) and anomaly areas related to geological boundaries (purple areas) were manually colored. In the figure, there are still noise anomaly areas that cannot be classified. In the future, it is expected that fault areas will be automatically extracted using a machine learning model that also uses fault distribution maps and geological maps as covariates.

6. Conclusions

Gamma-ray spectrometry is an excellent exploration method that can obtain information at a depth of 30 to 50 cm below the surface, which cannot be obtained by other remote sensing methods. The usefulness of this technology will be even greater if we can detect buried fault locations that cannot be directly observed due to recent sediments and soil. In order to clarify the possibility of fault detection by GRS, radioactivity prospecting including soil radon gas survey, etc., was carried out on already-known faults of four areas: active faults (the Adera fault, the Abashiri lake east coast fault group, and the Honguu fault) and non-active faults (Tanagura east marginal fault). As a result, it was shown that it is possible to detect fault traces with the following three indices: (1) Fault presumed from the continuity of anomaly points with an increasing rate Bi/Tl above the threshold defined for each study area. (2) Fault estimated from the continuity of peak points of non-diffusive radon. (3) Faults inferred from the continuity of conversion points where the average value changes in the stepped fluctuation pattern of nuclide and nuclide ratio. However, each index has an error of 0–30 m, 0–180 m, and 0–100 m from the published fault position, respectively. Published fault locations are compiled from literature maps and may have errors of up to 100 m or more depending on location. Therefore, it is concluded that the error of the fault position estimated by the three indices is within the allowable error range. Fault detection using radioactivity exploration should be based on the three indicators listed above. It is difficult to detect faults using only total gamma-ray measuring equipment such as survey meters, which has been done so far. In order to continue to develop analysis technology for radioactivity prospecting, in addition to accurate GRS and radon gas concentration measurements, it is also necessary to develop analysis technology using Artificial Intelligence (AI) technology, etc.

Author details

Imaizumi Masayuki
Nagasaki Techno Co., Ltd., Kochi City, Japan

*Address all correspondence to: imaizumi.masayuki@palette.plala.or.jp

References

[1] Hamachi T. World Uranium Resources. Vol. 119. Chishitsu News; 1964. pp. 1-13

[2] Suran J. Evaluation of effectiveness of uranium exploration methods (in Czech). Uhli, Rudy, Geologicky Pruzkum. 1998;**12**:387-389

[3] IAEA. Airborne Gamma Ray Spectrometer Surveying. Technical reports series No. 323. 1991

[4] IAEA. Guidelines for radioelement mapping using gamma ray spectrometry data. IAEA-TECDOC-1363. 2003

[5] Fortin R, Hovgaard J, Bates M. Airborne gamma-ray spectrometry in 2017. Solid ground for new development, airborne geophysics, paper 10. In: Tschirhart V, Thomas MD, editors. Proceedings of Exploration 17: Sixth Decennial International Conference on Mineral Exploration. Canada: Decennial Mineral Exploration Conferences; 2017. pp. 129-138

[6] Dickson BL, Scott KM. Interpretation of aerial gamma ray surveys-adding the geochemical factors. AGSO. Journal of Australian Geology & Geophysics. 1997; **17**(2):187-200

[7] Caciolli A, Baldoncini M, Bezzon GP, Broggini C, Buso GP, Callegari I, et al. A new FSA approach for in situ γ ray spectrometry. Science of the Total Environment. 2012;**414**:639-645. DOI: 10.1016/j.scitotenv.2011.10.071

[8] Ambronn R. Rn in soil gas. In: Jahrbuch des Halleschen Verbandes fur die Erforschung der Mitteldeutschen Bodenschatze. Vol. 3. No. 2. 1921. p. 21, 44

[9] Lane AC, Bennett WR. Location of a fault by radioactivity. Beitr. angewandten Geophysik. 1934;**4**:353-357

[10] Ochiai T. Prospecting of underground water by the surface detection of natural radioactivity. Journal of Geography, Tokyo Geographical Society. 1972;**81**(5):21-32 (in Japanese)

[11] Kimura S. Development of subsurface Geochemistric indications exhibited by ground and sea bottom gamma-rays. Radioisotopes. 1995;**44**: 627-636 (in Japanese)

[12] Imaizumi M, Takeuchi M, Komae T. The difference of radon concentration in soil-gas from gamma-ray spectrometry across active Adera fault in Yamaguti village, Nagano prefecture. Journal of the Japan Society of Engineering Geology. 1993;**34**(1):1-13 (in Japanese with English abstract)

[13] Ellis DV, Singer JM. Well Logging for Earth Scientists. 2nd ed. Dordrecht, The Netherlands: Springer; 2007. p. 692

[14] Fertl WH, Stapp WL, Vaello DB, Vercellino WC. Spectral gamma-ray logging in the Texas Austin chalk trend. Journal of Petroleum Technology. 1980; **32**:481-488

[15] Kimura S. A method for prospecting thermal water using surface distribution of gamma-rays coming from various nuclides. Journal of the Japanese Society of Hot Spring Sciences. 1987;**37**:73-92 (in Japanese)

[16] Imaizumi M. Engineering geological study on discontinuous deformation in rock mass using geochemical techniques. [PhD thesis] Chiba University. 1998; p. 204

[17] Imaizumi M, Komae T, Hamada H. Gamma-ray spectrometry for prospecting faults around Adera faults in

Yamaguchi Village, Nagano prefecture. Journal of the Japan Society of Engineering Geology. 1990;**33**(2):31-43 (in Japanese with English abstract)

[18] Imaizumi M, Tsuchihara T, Yoshimoto S, Ishida S, Uchiyama S, Yuuki Y. Radioactivity prospecting for groundwater development and a new prospecting device with a plastic scintillator. Jarq-Japan Agricultural Research Quarterly. 2010;**44**(3):291-299

[19] Szabó KZ, Jordan G, Petrik Á, Horvath Á, Szabó C. Spatial analysis of ambient gamma dose equivalent rate data by means of digital image processing techniques. Journal of Environmental Radioactivity. 2017;**166**:309-320

[20] Jolie E, Klinkmueller M, Moeck I, Bruhn D. Linking gas fluxes at Earth's surface with fracture zones in an active geothermal field. Geology. 2016;**44**(3): 187-190

[21] Yoshimura T, Matsumoto H. Basic problems and future issues of observation of fault zones by gamma-ray survey, Tsuchi-to-Kiso. Journal of Soil mechanics and foundation engineering, Japanese Geotechnical Society. 1994;**42** (5):41-46 (in Japanese)

[22] Reinhardta N, Herrmanna L. Gamma-ray spectrometry as versatile tool in soil science: A critical review. Journal of Plant Nutrition and Soil Science. 2018;**182**:9-27

[23] Katayama N. Review, Uranium and its Resources and Minerals. Vol. 1-12. Tokyo, Japan: Asakura Publishing Co., Ltd.; 1961. pp. 40-53

[24] Hatsuda Z. Distribution of Radioactivity in Granite, Uranium – Its Resource and Mineral. Tokyo, Japan: Asakura Publishing Co., Ltd.; 1961. pp. 183-189 (in Japanese)

[25] Nielson DL, Linpei C, Ward SH. Gamma-ray spectrometry and radon emanometry in environmental geophysics. In: Ward SH, editor. Geotechnical and Environmental Geophysics. Vol. 1. Houston, Texas, USA: Society of Exploration Geophysicists; 1990. pp. 219-251

[26] Stahr K et al. Beyond the horizons: Challenges and prospects for soil science and soil Care in Southeast Asia. In: Fröhlich HL, Schreinemachers P, Stahr K, Clemens G, editors. Sustainable Land Use and Rural Development in Southeast Asia: Innovations and Policies for Mountainous Areas. Springer Environmental Science and Engineering. Berlin, Heidelberg: Springer; 2013

[27] Adams JAS, Weaver CE. Thorium to uranium ratios as indications of sedimentary processes: Example of concept of geochemical facies. American Association of Petroleum Geologists Bulletin. 1958;**42**:387-430

[28] Dudek L, Klaja J. Geological interpretation of spectral gamma ray (SGR) logging in selected boreholes. Nafta-Gaz. 2016;**1**:3-14

[29] Schlumberger. Log Interpretation Charts. New York, USA: Schlumberger; 1985. p. 207

[30] Nuțu-Dragomir ML, Chitea F, Stochici R, Diacopolos C. A new approach concerning active faults in Subcarpathian nappe (EAST CARPATHIANS). In: Geoscience; Second Symposium of SGAR. Romanian: Romanian Society of applied Geophysics (SGAR); 2017

[31] Tanaka K. Evaluation of fault activity by characteristics of intrafault materials, proceeding of symposium on investigation method and interpretation of active fault. In: Symposium

Proceedings, Japan Society of Engineering Geology. 1993. pp. 15-26 (in Japanese)

[32] Kanamori Y, Inohara Y, Miyakoshi K, Satake Y. Characteristics of intrafault materials within the Atotsugawa fault of Central Japan (part 1). Journal of the Japan Society of Engineering Geology. 1982;**1.23, 3**:137-155 (in Japanese with English abstract)

[33] Latham AG, Schwarcz HP. Review of the modelling of radionuclide transport from U-series disequilibria and of its use in assessing the safe disposal of nuclear waste in crystalline rock. Applied Geochemistry. 1989;**4**:527-537

[34] Smellie JAT, Mackenzie AB, Scott RD. An analogue validation study of natural radionuclide migration in crystalline rocks using uranium-series disequilibrium studies. Chemical Geology. 1986;**55**:233-254

[35] Japanese Geotechnical Society. Method for Engineering Classification of Rock Mass, JGS:3811-2004. Tokyo: Japanese Geotechnical Society; 2004. p. 46

[36] McQueen KG. Identifying Geochemical Anomalies, Project, Regolith Geology and Mineral Exploration. Australia: University of Canberra; 2008. pp. 1-7

[37] Reimann C et al. Background and threshold: Critical comparison of methods of determination. Science of the Total Environment. 2005;**346**:1-16

[38] Active Fault Society. Newly-Edited Active Fault in Japan – Its Distribution Map and Data. Tokyo, Japan: University of Tokyo Press; 1991. p. 437 (in Japanese)

[39] Yamada N. Trace of Adera fault (part 1) - from tunnel in Mt. Ena to Tsukechi. Geologic News. 1978;**238**: 37-49 (in Japanese)

[40] Chiba T, Suzuki Y. Red relief image map: New visualization method for three-dimensional data. Japan Society of Photogrammetry and Remote Sensing. 2004;**15**:81-89 (in Japanese)

[41] Kinki Regional Agricultural Administration Office. Groundwater Investigation for Agriculture - Report of Regional Development Investigation in Kii-Tanabe District, Resource Section. Kyoto, Japan: Kinki Regional Agricultural Administration Office; 1993. p. 170 (in Japanese)

[42] Tohoku Regional Agricultural Administration Office. Study on groundwater of granitic rock body in the southern Abukuma plateau, Japan. In: Report on Development of Groundwater. Department of Planning, Tohoku Agricultural Administration Bureau: Southern Abukuma District. 1990. p. 206 (in Japanese)

[43] Koshiya S, Nagae R, Okami K, Morita Y. Fracture system and ground water of granitic rock body in the southern Abukuma plateau, Japan. Journal of the Japan Society of Engineering Geology. 1991;**32**(4):23-39 (in Japanese with English abstract)

[44] Tanaka H, Ochiai K. Granitic rocks accompanied with pegmatites containing rare element minerals in the Uzumine area, central Abukuma Mountains. Journal of Mineralogical and Petrological Sciences, Japan Association of Mineralogical Sciences. 1989;**83**:318-331 (in Japanese with English abstract)

Section 2

Gamma Rays in Agriculture

Chapter 4

Role of Gamma Irradiation in Enhancement of Nutrition and Flavor Quality of Soybean

Kalpana Tewari, Mahipal Singh Kesawat, Vinod Kumar, Chirag Maheshwari, Veda Krishnan, Sneh Narwal, Sweta Kumari, Anil Dahuja, Santosh Kumar and Swati Manohar

Abstract

Soybean has the potential to be termed the "crop of the future" due to its significant capacity to address protein-energy malnutrition and hidden hunger, particularly in developing countries where diets are predominantly based on wheat and rice. Despite its substantial nutritional value, numerous health benefits, and its versatility in various food and industrial applications, soybean's full potential remains underutilized due to inherent off-flavors and the presence of antinutritional factors (ANFs). Gamma irradiation is known to have a positive impact by inducing structural and chemical changes in biomolecules like carbohydrates, lipids, proteins, and other phytochemicals. This process leads to improved functionality and market demand by reducing ANFs and the off-flavor in soybeans. Scientifically, it has been demonstrated that low to moderate doses of gamma radiation, up to 10 kGy, can positively influence the antioxidant capacity of soybeans. This, in turn, helps control lipid and protein oxidation, reducing the generation of off-flavors and enhancing the quality and nutraceutical potential of soybeans.

Keywords: γ-Irradiation, soybean, antinutrional factors, nutrition, off-flavor, food quality, antioxidant potential, allergen

1. Introduction

Soybeans hold the position of being the most widely cultivated oilseed crop, both in India and across the world. According to data from Our World in Data in 2021, soybeans were grown on around 129.52 million hectares of land globally. In 2021, global soybean production reached 371.69 million metric tons, and there are expectations that this figure will rise to 391.17 million metric tons by 2023. The major countries contributing to soybean production are the United States, Brazil, Argentina, China, and India.

Soybean seeds have a composition comprising roughly 27% complex carbohydrates, 40% protein, 20% oil, 8% moisture, and 5% minerals. This nutritional profile positions soybean as a valuable resource for addressing malnutrition and undernutrition, particularly in developing countries. Soybean's protein quality, measured by the protein digestibility corrected amino acid score, is on par with protein sources derived from animals.

Irradiation involves the use of radiation on foods primarily for extended hygiene and safety purposes and others too. It is a low-cost, environment-friendly, non-thermal physical processing technology widely used in the food industry. It has been reported to reduce food allergenicity, effectively killing insects, molds, and bacteria [1, 2] and also to minimize the harmful substances such as biogenic amines [3] and anti-nutritional factors present in food [4]. Besides, when applied in the proper doses, it has no detrimental effect on the physicochemical, nutritional, and sensorial quality of food [5–7].

Irradiation is categorized into two types: ionizing radiation, which includes UV-C, γ-irradiation, electron beams, and X-rays, and non-ionizing radiation, which encompasses ultraviolet, visible light, and infrared. Ionizing radiation possesses higher energy levels compared to non-ionizing radiation. Among these irradiation methods, gamma irradiation is the most commonly used due to its exceptional penetration capabilities, as highlighted in a study by Liu et al. [8].

Generally, electron beam irradiation can only penetrate the surface of food to a depth of 50–150 mm, as noted by Fan et al. [9], while γ-irradiation can deeply and readily penetrate various materials, as mentioned by Ekezie et al. [10]. In industrial applications, the two primary sources of gamma radiation are cobalt-60 (^{60}Co) and iridium-192 (^{192}Ir). The energy of ionizing radiation is measured in electronvolts (eV), but due to the small size of an electronvolt, it's common to express ionizing radiation energy in megaelectronvolts (MeV). Co-60 emits two γ-rays simultaneously with energies of approximately 1.17 and 1.33 MeV during its radioactive decay process. Additionally, ^{60}Co emits a low-energy electron (beta particle) with a maximum energy of around 0.3 MeV [9]. ^{192}Ir emits gamma rays at energies of 0.31, 0.47, and 0.60 MeV. Given that the radiation from ^{60}Co has roughly twice the energy of that from ^{192}Ir and possesses greater penetration power through materials, making it more hazardous and requiring more substantial shielding. The standard unit for measuring the absorbed dose of γ-radiation is the "rad" (Radiation Absorbed Dose). It is defined as a dose of 100 ergs of energy per gram of the given material. The International System of Units (SI) unit for measuring absorbed radiation is the gray (Gy), defined as joules per kilogram of mass. One gray is equivalent to 100 rads.

The impact of γ-irradiation on food depends on whether it is in a dry or aqueous state. In solid foods, the food molecules directly absorb radiation energy, leading to structural changes. In the case of aqueous foods, when exposed to γ-radiation, water generates hydroxyl radicals and hydrated electrons, which subsequently react with biomolecules to form covalent bonds [11]. Proteins in soybeans are influenced by γ-irradiation through mechanisms such as conformational changes [12] and the promotion of amino acid oxidation, peptide bond cleavage, and the formation of disulfide bonds [13]. γ-irradiation has a long history as a food preservation method. Currently, it is widely acknowledged that foods irradiated at doses below 10 kGy are safe and pose no toxicity risks to humans [7, 14]. The Food and Agriculture Organization (FAO) and the World Health Organization (WHO) have also stated that the maximum absorbed dose delivered to foods should not exceed 10 kGy, except when it is necessary to achieve a legitimate technological purpose [15]. However, it's important to note that the specific irradiation dose can vary based on the type of food, its intended use, and regulations in different countries. For instance, in the United States, the irradiation dose for dry spices/seasonings is 30 kGy, and for frozen packaged meats, it is 44 kGy.

Hence, the impact of γ-irradiation on the nutritional aspects and flavor of soybean seeds and their processed products has been extensively explored in the literature. This study could potentially serve as a valuable resource for preserving the nutritional value of soybeans and enhancing their bioavailability and scientific acceptance to its worldwide consumers.

2. Impact of γ-irradiation processing on inherent properties of soybean

Ionizing radiation is a highly efficient method for controlling pathogens, preventing spoilage, and enhancing the storage life of various food products, including spices, herbs, fresh fruits, vegetables, leguminous seeds, and cereals. Research conducted by the World Health Organization has confirmed that food irradiation does not raise any significant concerns regarding toxicity, microbiology, or nutritional content [16]. However, it does induce chemical changes that may impact the overall quality of the products. It is generally considered safe and nutritionally sound to consume food irradiated at doses up to 10 kGy, with some products even approved for higher doses [17].

The effects of γ-irradiation on various inherent properties of soybeans are summarized in **Figure 1** and extensively discussed in subsequent sections.

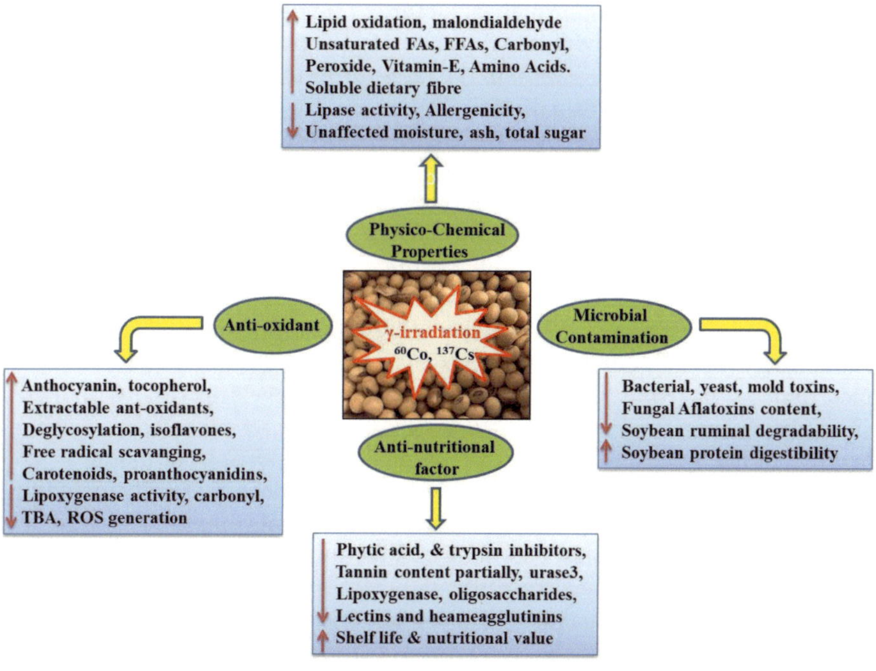

Figure 1.
Effects of γ-irradiation on soybean. FAs: Fatty acids, FFAs: Free fatty acids, ROS: Reactive oxygen species, TAB: Thiobarbituric acid, Co- cobalt, Cs- Cesium. Recommended dose of γ-irradiation is 0.5–10 kGy. Down and up arrow represent decrease and increase respectively.

2.1 Physicochemical profile

Exposing soybean to γ-irradiation levels of up to 10 kGy has been observed to effectively preserve essential nutritional aspects. Soybean oil extracted from soybeans subjected to gamma radiation up to 10 kGy exhibited no significant changes in various physicochemical characteristics, including lipid content, fatty acid composition, acid value, peroxide value, and trans fatty acid content [18]. However, there was a clear trend of prolonged induction periods with higher radiation doses. Beyond the 10 kGy thresholds, a notable increase in n-hexanal content was detected, indicating its potential as an indicator of excessive radiation exposure in soybeans. Additionally, as the irradiation dose increases, there is a tendency towards an extended period of resistance to oxidation, which is known as the "induction period." However, it's worth noting that when irradiation doses exceed 10 kGy, there is a noticeable increase in the levels of n-hexanal, which can potentially serve as an indicator of excessive irradiation in soybeans. In the case of peanut seeds, an irradiation dose of 1 kGy is considered suitable. Nevertheless, a higher dose of 10 kGy has been found to significantly reduce the fat and protein content of peanuts while increasing their water activity. On the other hand, moisture, ash, and total sugar contents are unaffected by γ-irradiation [8]. γ-irradiation accelerates lipid oxidation, leading to increased values for fatty acid, peroxide, carbonyl, and malondialdehyde, while reducing lipase activity. Additionally, γ-irradiation induces changes in fatty acid and amino acid composition. Hazelnut kernels irradiated with a 0.5 kGy dose maintain high quality in terms of free fatty acids, peroxide content, and vitamin E [19]. The impact of γ-irradiation on protein properties and structures was examined, with doses up to 10 kGy tested. [20], studied the impact of mildew and gamma irradiation (using ^{60}Co-γ rays) on the digestibility, composition, and protein structure of soybeans. Mildew reduced pepsin digestibility and amino acid content while altering the protein's secondary structure. In fresh soybeans, ^{60}Co-γ irradiation in the range of 10 to 30 kGy had no significant effect on amino acids and pepsin digestibility, except for changes in protein secondary structure. However, in mildew soybeans, γ-irradiation reduced amino acid content and caused structural changes in the protein. As the radiation dose increased, there was a partial reduction in protein concentration, resulting in decreased solubility in water. This could be attributed to alterations in protein conformation, potentially causing hydrophobic amino acid residues originally hidden within the protein to become exposed on its surface, thus increasing the protein's hydrophobic nature. The irradiated samples displayed diminished emulsifying abilities and heightened surface hydrophobicity [21–23]. Furthermore, this dose does not adversely affect the sensory characteristics of hazelnut kernels. However, vitamin E levels decrease with increasing irradiation doses.

In oil seeds like soybean, peanut, and sesame, the proportions of induced unsaturated fatty acids are dependent on the irradiation dose, generally increasing with higher doses for different seed types [24]. The natural compounds found in soybeans, including isoflavones, Bowman-Birk factor, tocopherols, lecithin, lunasin, saponins, and others, have shown significant health benefits in combating conditions like cancer, atherosclerosis, diabetes, and osteoporosis, as evidenced by Kumar et al. [25] and Messina [26]. This has led to soybeans being recognized as a "functional food." However, the global consumption of soybeans as a dietary source remains low, with only around 10% of soybean production being used as food, primarily in East and South East Asia. One of the reasons for this limited consumption is the presence of anti-nutritional factors like phytic acid, Kunitz trypsin inhibitor, and allergens, including Gly m Bd 68 K, a subunit of β-conglycinin, which can deter people from incorporating soybeans into their diets. Meinlschmidt et al. [27, 28] conducted research on the impact of γ-irradiation (at doses of 3–100 kGy) on the immunogenecity of soybean protein isolate and observed a remarkable reduction of 91–100% in the soluble protein fraction's allergenicity. γ-irradiation achieves this by introducing structural changes, promoting protein aggregation or degradation, and destroying exposed epitopes [29]. The aversion to including soybeans in one's diet is often linked not only to allergens but also to the characteristic beany flavor associated with soybean products. This off-flavor is primarily generated through three processes: automatic, photosensitized, and enzymatic oxidations. Among these, enzymatic degradation, specifically the breakdown of unsaturated fatty acids catalyzed by lipoxygenases (LOX) in soybean seeds, is the primary pathway for producing compounds responsible for the beany flavor [30]. This process involves the catalytic oxidation of polyunsaturated fatty acids containing 1,4-Z, Z-pentadiene structures, such as linoleic and α-linolenic acid, by soybean seed lipoxygenases. The resulting hydroperoxides are then further metabolized by hydroperoxide lyases, leading to the production of volatile aldehydes and ketones that give soy products their distinctive off-beany flavor [21, 22]. The beany flavor of soybean is associated with more than 20 volatile compounds, including fatty aldehydes, fatty alcohols, fatty ketones, furans, furan derivatives, and aromatic compounds [30]. Of these, hexanal and (Z)-3-hexenal are the most significant contributors [31]. Hexanal has a scent reminiscent of cut grass and green notes, with an incredibly low odor threshold of 0.0045 ppb, and it results from the enzymatic oxidation of linoleic acid. (Z)-3-Hexenal, with a green beany flavor, also strongly influences the soybean's overall flavor profile and has an extraordinarily low odor threshold of 0.00012 ppb. It is formed through the oxidation of α-linolenic acid [32]. Apart from these volatile compounds, certain non-volatile compounds like flavonoids, saponins, phenolic acids, and specific amino acids can also contribute to the formation of the beany odorants [30]. The content of soluble dietary fiber in soybean increases when exposed to higher irradiation doses, ranging from 400 kGy to 1200 kGy. The combination of γ-irradiation and micronization (grinding to produce soft mass/powder) demonstrates the most effective means of enhancing dietary fiber degradation while also positively influencing its physical and chemical characteristics [33, 34]. As a result, this method is considered an optimal approach for enhancing the quality of soybean overall nutrition. These investigations collectively underscore the promise and potential utility of γ-irradiation as a technology for maintaining and enhancing the physicochemical properties and nutritional value of soybean.

2.2 Antioxidants

The antioxidant properties of soybeans can be significantly enhanced through γ-irradiation. The seed coat color has a prominent effect on the antioxidant response to γ-radiation. In a study conducted by Tewari and colleagues [35], soybeans were exposed to low doses of gamma rays (0.25 kGy, 0.5 kGy, 0.75 kGy, and 1.0 kGy). The findings indicated that soybean varieties with darker seed coat colors displayed greater susceptibility to γ-irradiation. A significant observation was the marked rise in the levels of tocopherols and anthocyanins in dark-colored soybean varieties. Further, at a radiation dose of 0.25 kGy, there was a notable increase in the concentration of isoflavones. Additionally, γ-radiation had a significant reducing effect on parameters associated with undesirable flavors, including seed lipoxygenase activity, thiobarbituric acid number, and carbonyl value in the dark-colored soybean varieties.

Radiation dose appears to play a crucial role in the extractable phenolics found in legume and cereal seeds, which may also influence the antioxidant properties of soybeans. However, it's essential to note that γ-irradiation is not currently employed to intentionally boost the antioxidant properties of soybeans. Nevertheless, under certain favorable radiation conditions, an increase in the content of extractable phenolics has been detected. Kumari and co-workers [36], conducted a study in which γ-irradiation resulted in a decrease in non-extractable phenolics in both yellow and black-coated soybean cultivars. Remarkably, a substantial increase in the total extractable flavonoids was noted in irradiated samples, which included soybeans [37], peanut skins [38], and quinoa subjected to electron beam irradiation [39]. On the contrary, some researchers have reported a significant enhancement in anthocyanin content in pigmented rice and soybeans as a result of γ-irradiation Taken together, these findings imply that γ-irradiation can impact the antioxidant properties of soybeans and other food products. However, the precise effects may vary depending on factors such as the type of seed, radiation dose, and other environmental conditions. Soybeans are rich in isoflavones, and their levels are influenced by genetic factors. Isoflavones, a type of polyphenol, serve as antioxidants that can help prevent osteoporosis, reduce the risk of atherosclerosis, and protect against cardiovascular disease. The impact of γ-irradiation on isoflavones has been predominantly studied in soybeans. The specific isoflavones, like genistein, daidzein, and glycetein, showed a significant increase in concentration during irradiation processing up to 10 kGy. Additionally, measures of antioxidant potential, such as DPPH (1,1-diphenyl-2-picrylhydrazyl) scavenging activity and enhanced hydroxyl radical scavenging, exhibited a positive correlation with the radiation dose, suggesting that using radiation as a food preservation method can have beneficial nutritional implications [37, 40, 41]. However, at 50 kGy, isoflavone levels in the samples showed a lower, though not statistically significant change [42]. Interestingly, aglycons and isoflavone glucosides have displayed diverse responses to γ-irradiation. Variyar and group [41] noted that γ-irradiation (0.5–5 kGy) remarkably reduced the overall content of isoflavones, particularly the glycosides, while enhancing the content of isoflavone aglycons in 80% methanol extracts of soybeans. However, there have been reports where γ-irradiation treatments had no significant impact on the isoflavone content of soybeans when the radiation dose was below 10 kGy [40]. This variation could potentially be linked to the choice of solvent utilized for extraction. Furthermore, reduction in the overall isoflavone content was solely noted in yellow soybeans subjected to γ-irradiation at a dose of 1 kGy. This decline in non-extractable phenolics can be linked to the lower presence of protective anthocyanins, which serve as guards against radiolytic degradation, particularly when compared to black soybeans [43]. Among black soybeans, the metal chelating (Fe^{2+}) activity and reducing power exhibited a decreased sensitivity to γ-irradiation. Conversely, yellow and green-coated soybeans demonstrated a notable enhancement in these attributes at radiation doses of 0.5 and 2 kGy. Moreover, Krishnan et al. [44] and Bansal et al. [45] reported that an effective improvement in extractable antioxidant activities and the preservation of nutraceutical properties was achieved with the application of a lower dose, specifically 0.5 kGy.

Radiation-induced transformations within the same class of phenolic compounds, such as the conversion of glycosides to aglycons, are largely attributed to the breaking of glycosidic bonds—a process known as deglycosylation. For example, γ-irradiation of soybeans has been shown to increase isoflavone aglycons and decrease their glucosides due to the deglycosylation of isoflavone glucosides [41]. They also showed that the increased DPPH scavenging activity of isoflavones in γ-irradiated soybeans was attributed to higher levels of the free form, such as isoflavone aglycon, despite a reduction in both the total isoflavone content and its glucosides. The hydroxyl radical scavenging activity, as evidenced by its ability to inhibit deoxyribose oxidation, stayed at a high level up to a radiation dose of 2.0 kGy in soybeans. This suggests that γ-irradiation contributes to an increase of antioxidant activity. Medium doses of gamma irradiation (1–10 kGy) were reported to enhance total phenolic and tannin contents, as well as DPPH scavenger activity, in soybean seeds, while reducing protein oxidation. Although gamma irradiation had minimal impact on lipid peroxidation and soluble protein content, a significant decrease in protein oxidation was observed at a 10 kGy dose, suggesting improved antioxidant activity depending of radiation dose [46]. In another study, Mata-Ramírez and group [47] conducted an assessment of the quantification and bioactivity of isoflavones in soybean callus subjected to UV-light-induced stress. Their results suggested that exposure to UV-C light stress resulted in elevated levels of genistein-O-glucosyl-malonate and genistein-O-glucoside in soybean callus, leading to improvements in both antioxidant and anti-inflammatory activities [48]. This suggests that the application of both tissue culture techniques and UV-B radiation treatment can be an effective means of obtaining bioactive compounds. The increased antioxidant activity observed in soybean genotypes following exposure to low doses of gamma irradiation may be associated with the production of free flavonoid. Research has shown that these liberated flavonoids display stronger antioxidant characteristics in comparison to glycosides. Variyar et al. [41] proposed that the radiation-induced degradation of glycosides leads to the liberation of free isoflavones. Given that isoflavones are a type of phenolic compound, the rise in their concentration at lower doses of gamma irradiation corresponds with the concurrent increase in total phenolic content. This, in turn, contributes to the improvement in antioxidant activity.

The improvement in soybean antioxidant activity was linked to an increase in the overall phenolic content. It was proposed that γ-irradiation could boost phenolic content by stimulating the activity of phenylpropanoid pathway enzymes. The increase in isoflavones was theorized to result from either the conversion of malonyl derivatives into free glycosides or an augmented biosynthesis. Using the epithelial cell line BEAS-2B, [43] explored the effect of three different doses of γ-radiation, for instance 0.25, 0.5 and 1.0 kGy, on the anti-proliferative and cytoprotective effect of black and yellow soybean extracts that were characterized in terms of total phenolics, and flavonoids and the anthocyanin cyanidin 3- glucoside. It was reported that γ-irradiation at 1 kGy dose enhanced the cyanidin 3- glucoside content by 33%, which correlated well with a 78% decrease in reactive oxygen species (ROS). Also, irradiation up to 0.5 kGy enhanced the total phenolic content in black and yellow soybeans. The flavonoids, particularly daidzein, showed an increase at radiation doses of 0.25, 0.5, and 1.0 kGy in black soybeans, and at 0.25 and 0.5 kGy in yellow soybeans. The rise in total phenolic and anthocyanin content in irradiated soybeans was attributed to an enhancement in the activity of enzymes like flavonoid glucosyltransferase and phenylalanine ammonia-lyase or to the cleavage of glycosidic linkages in bound procyanidins, leading to their free monomeric form.

Medium doses of γ-radiation have been shown to have a positive effect on the antioxidant potential of seeds that, in turn, combats the free radical-induced lipid and protein oxidation, thus minimizing off-flavor production [35, 44, 49]. Kumari and co-workers [36] revealed that γ-irradiation at 2.0 kGy dose improved the protein quality of soybeans by reducing protein oxidation and increasing protein solubility. The report indicated a 37% reduction in protein oxidation, as measured by the carbonyl number, in black soybeans and a 28% reduction in yellow soybeans. They noted that increasing the radiation dose to 5 kGy resulted in elevated protein oxidation levels, similar to those of untreated soybeans. Further, was noted that γ-irradiation up to a 5 kGy dose reduced the activity of LOX isozymes, which improved protein quality by limiting interactions between lipid hydroperoxides and proteins. Additionally, γ-irradiation up to a 5 kGy dose increased free radical scavenging activity, as assessed through the DPPH assay. However, the FRAP values increased up to a radiation dose of 2 kGy and decreased beyond that point.

Tocopherols, found abundantly in plant-derived lipids, are the most prevalent natural antioxidants, acting as sacrificial guardians for the lipid components within seeds. In the autoxidation process, they function as chain-breaking antioxidants by capturing free radicals, thus effectively thwarting the initiation and halting the propagation of oxidation. Braunrath et al. [50] observed significant reductions in peroxides and hydroperoxides when δ- and γ-tocopherols were present. Contrastingly, Yun et al. [51] noted that soybeans treated with γ-irradiation at doses of 5.0 and 10.0 kGy displayed a substantial decline in α-tocopherol, with decreases of 24.3% and 35.4%, respectively. However, Dixit et al. [52] discovered that there were no alterations in the overall tocopherol content in soybeans across the entire range of γ-irradiation doses, which included values from 0.5 to 5 kGy. The decrease in tocopherol content in irradiated oil is believed to be associated with the degradation of these antioxidant compounds that take place during the irradiation process.

Minami and co-workers [53] have shown that γ-irradiation, even up to 80 kGy, had no impact on the quantity of unsaturated or saturated fatty acids in soybeans. This phenomenon is attributed to the ample presence of antioxidants, such as tocopherols and carotenoids, in soybeans. These antioxidants effectively scavenge the radicals generated by γ-irradiation. Similarly, in a study by Hafez and colleagues [54], it was noted that no alterations were detected in the fatty acid composition, which included C16:0, C18:0, C18:1, and C18:2, in soybean oil. This observation remained consistent even when exposed to relatively high doses of γ-irradiation, up to 100 kGy. However, a negative correlation was established between the concentration of linolenic acid and the applied radiation dose. Aziz et al. [55] observed that γ-irradiation led to a reduction in the acid values of cereal fats, with the effect becoming more pronounced with the increment in radiation dose within the range of 5–15 kGy. The decrease in acid values can be ascribed to the formation of free radicals during irradiation, which impacts the unsaturated fatty acids in the oil, leading to the production of hydroperoxides and peroxides, as described by Mahrous [56].

Mexis and Kontominas [57] and Bhatti and group [58] found a significant increase in the peroxide value of oils obtained from irradiated almonds and peanuts as the γ-irradiation dose increased. The decrease in the iodine value of oils can be connected to the formation of peroxide compounds and the saturation of double bonds in unsaturated fatty acids, induced by irradiation. Bhatti and co-workers [59] noted a noticeable decline in the iodine value for γ-irradiated peanuts.

2.3 Anti-nutritional factors

Leguminous seeds' protein content makes up 20% of the total plant-based protein consumption in humans [60]. Nonetheless, specific legumes, particularly soybeans, contain substantial amounts of bioactive compounds that can modify the way nutrients are processed in the body upon consumption. The primary proteins that contribute to the reduction in the nutritional value of unprocessed soybeans are lectin and trypsin inhibitors. However, it's important to acknowledge that other naturally occurring compounds can also contribute to the observed adverse effects. To mitigate losses during the harvest and storage of these grains, the use of ionizing radiation presents itself as an appealing and health-conscious alternative when compared to conventional chemical treatments. The application of ionizing radiation for the purpose of preserving and sanitizing grains offers a promising strategy to extend shelf life and minimize losses during storage. The anticipated costs and benefits associated with commercial radiation treatment appear to be competitive when compared to other fumigation techniques, as well as thermal and physical treatment methods. Abu-Tarboush [61] reported a 34.9% reduction in trypsin inhibitory activity in soybean flour was assessed after exposure to a γ-radiation dose of 10 kGy. Similarly, Farag [62] found that as radiation doses increased (5, 15, 30, and 60 kGy), with the increment in radiation dose, the degree of inactivation of trypsin inhibitory activity also rose, leading to losses of 41.8%, 56.3%, 62.7%, and 72.5%, respectively. In the research conducted by Villavicencio et al. [63], it was noted that in the Carioca variety, the tannin concentration remained unchanged during soaking and cooking, while in the Macacar variety, it showed an increase. In contrast, Mechi et al. [64] reported a decline in tannin content in black beans when subjected to γ-radiation combined with the cooking process. Amaya et al. [65] proposed that the decrease in tannin content due to cooking is evident and is linked to alterations in chemical and solubility reactivity. This decrease can be attributed to various factors, such as interactions with organic substances, proteins, and modifications in their chemical structure, as suggested by Brigide and Canniatti-Brazaca [66]. Furthermore, it was observed that γ-radiation resulted in a reduction in tannin content as the radiation dose increased, up to a certain threshold. This decrease in tannin content is particularly advantageous, as tannins are recognized as antinutritional factors that can hinder protein digestibility [63, 64]. When the tannin-to-protein ratio reaches 5:1, all the protein tends to precipitate due to the action of tannins, as explained by Pino and Lajolo [67]. Therefore, these findings indicate that, concerning antinutritional factors, their levels decreased with higher radiation doses in cooked and raw samples, with the cooking process itself also contributing to the decrease in these antinutritional factors.

γ-irradiation has been demonstrated to enhance the nutritional quality of soybeans by reducing anti-nutritional components, including phytic acid, trypsin inhibitors, lipoxygenase, oligosaccharides, tannins, urease 3, and haemagglutinating agents, [62, 68, 69]. Currently, there are over 51 countries, China included, have officially approved the use of γ-irradiation in food processing, typically at an average dosage of 10 kGy [70]. A joint study group by FAO/IAEA/WHO has reported that γ-irradiation at doses exceeding 10 kGy remains safe and maintains nutritional adequacy [6]. Therefore, it is plausible that the application of high-dose γ-irradiation could serve as an effective approach to improve the quality and safety of fungus-contaminated soybeans intended for animal feed. Soybean seeds inherently contain three lipoxygenase enzymes, which can impart undesirable beany or grassy flavors. The elimination of these lipoxygenases is a method to enhance the stability and taste of soybean oil and protein products. Additionally, the presence of phytic acid and its mineral derivatives, phytates, creates nutritional and environmental challenges due to their limited digestibility in monogastric organisms. As a result, there has been a substantial rise in endeavors to grow crops with lower phytic acid content. Soybean varieties created through γ-radiation-induced mutations have shown enhanced nutritional qualities owing to their reduced phytate levels along with the reduced lipoxygenase enzyme activity in their seeds [71, 72].

2.4 Allergen

Soybean is a protein-rich food (besides having nutritious fiber, fatty acids, vitamin, minerals etc.) consumed globally. Additionally, it also contains anti-nutritional factors, like agglutinin, protease inhibitors, and allergenic proteins that elicit adverse immune responses, which significantly hamper its nutritional and functional profits. Most of the allergens in soybean are proteins and more than 33 allergens have been characterized so far [73], which influenced around ~0.5% of the general population globally [74]. It affects 0.4% children in USA [75] and its mild symptoms usually appear up to 2 years, however, severe immune responses may persist the whole life [76]. Soyben-induced allergic reactions (allergenicity) have a wide impact on the skin (dermatitis, eczema, urticaria, acne etc) respiratory system (asthma, bronchospasm, dyspnea, edema, Rhinitis wheezing etc.), gastrointestinal tract (colitis, enterocolitis, bowel disease, vomiting etc.) and other prominent symptoms like conjunctivitis, lethargy etc. [76]. Soybean is 8th most common food allergens and become a public health problem globally for its manufacturers as well as consumers [77, 78]. In soyaben allergic patients, its value ranges from 0.0013 to 500 mg, provoking mild to severe immunogenic response [79, 80]. So far, various allergens belonging to the cupin super-family, [81], have been identified from molecular weight ranging from 7 kDa (Gly m1a, Hull protein) to 75 kDa (α' Subunit of β-conglycinin) [76], provoking immunogenic responses in human [79] as well as various animals [82]. Among them, Glycinin and β-conglycinin are widely studied allergens [82, 83].

Major allergenic proteins include Gly m 1 to Gly m 8, Gly m Bd 28 K(P28), Gly m Bd 30 K(P34), [84, 85], Gly m Bd 60 K [86], and the Kunitz soybean trypsin inhibitor (KSTI, 20 kDa), [87] were detected in soybean allergic patients. Gly m 1 (~ 8.3 kDa, hydrophobic protein) and Gly m 2 (~ 8 kDa, defense protein) are considered as hull proteins [88] due to their location in soybean hull and accounts for their 75% and 25% allergenicity respectively [89]. Gly m 3 (~ 14 kDa, hydrophilic protein) is from profilin and labile to heat and hydrolysis [90]. Gly m 4 (~17 kDa, hydrophilic, and pathogenesis related protein), is susceptible to heat, acid, proteases, and found as protein isolates [91]. P34, P28 and Gly m 5 (β-conglycinin, ~ 150 to 210 kDa) are derivatives of 7S globulin and constitutes ~30%, Gly m 6 is a hexameric protein originate from 11S globulin, comprises ~40%, [92, 93], Gly m 7 (~ 76.2 kDa, biotinylated protein) is abundant in seed embryo, heat stable, and comprises <0.01% [75], and Gly m 8 (~ 28 kDa) derivative of 2S albumin comprises 10% of the total soybean protein [94]. Furthermore, Gly m Bd, *P34* (~ 30 kDa, anchorage vacuole protein) comprises ~1% [85, 95], and Gly m Bd, P28, (~28 kDa, glycosylated protein) comprises <0.5% of total seed protein [95, 96]. Soybean allergenicity is an IgE-mediated immune response, devoid of immunologic and clinical lenience resulting in clinical symptoms and pathological manifestations [97]. In continuation, Gly m 4, Gly m 5 and Gly m 6 provokes robust immune response in soybean-allergic patients [98]. Gly m 5 and Gly m 6 soybean allergens comprising 30% and 40% of the total soybean protein respectively [99] and are easily identifiable in almost every pediatric patient [100]. Schematic representation analyzing impact of low and high doses of y-irradiation on various pathophysiological aspects of soybean allergenicity, nutrients and intrinsic physiochemical properties is summarized in **Figure 2**.

γ-irradiation are primarily used for increasing food shelf-life by abolishing the surface pathogens; however, variable doses and combination of other factors later resulted in augmented additional microbes destruction and marked decrease in protein loss during food storage as well as reduced allergenicity [101]. Food exposure to y-irradiation

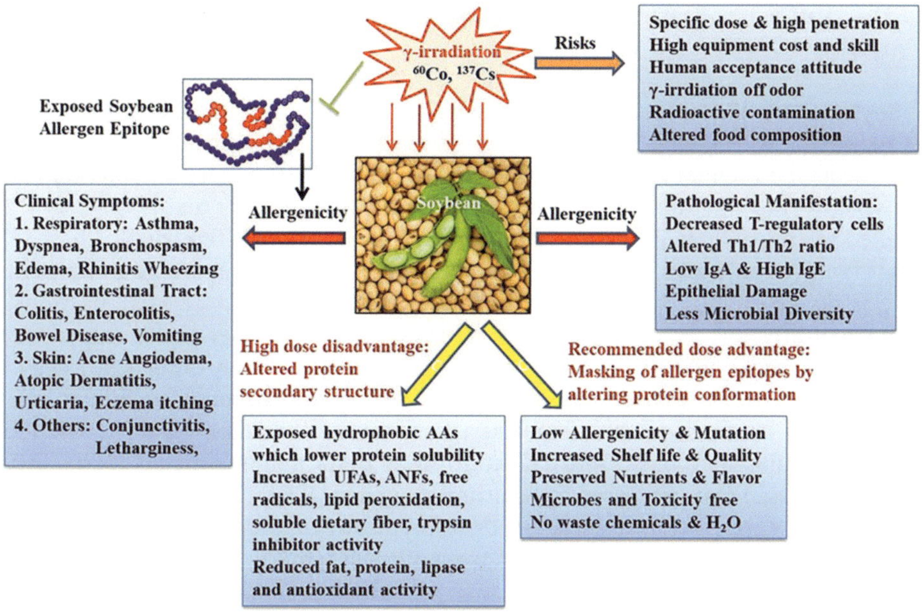

Figure 2.
Schematic representation of various pathophysiological aspects of soybean allergen, and risks as well as benefits associated with the use of γ-irradiation on intrinsic properties of soybean. Cobalt-60 (^{60}Co) and Cesium-137 (^{137}Cs) are major radioactive atoms producing gamma radiation, which alters the conformation of soybean allergen protein and, thus, masks the exposed antigenic epitope responsible for its augmented allergenicity contributing to clinical symptoms and pathological manifestations healthwise. The recommended dose of γ-irradiation is 0.5–10 kGy masks the soybean allergenicity, and microbial infection as well as increases the shelf life and nutrient quality. The higher dose of γ-irradiation (10 kGy–400 kGy) induces a change in secondary structure provoking several disadvantages by altering soybean nutrients, constituents, and various intrinsic properties. AAs: Amino acids UFAs: Unsaturated fatty acids, ANFs: Anti-nutritional factors.

involves production of low energy electron (β-particle) below 5 MeV from radioactive atoms ^{60}Co or ^{137}Cs, during radioactive decay [9], and modulates protein conformation structurally, resulted in masking of exposed epitopes in allergen protein, thus, reduced adverse immune response or allergenicity [10]. However, factors affecting reduced allergenicity depend on its types, exposure times and doses. ^{60}Co involves simultaneous generation of two γ-rays (1.17 and 1.33 MeV) and low energy β-particle (0.3 MeV), while ^{137}Se produces γ-rays of energy 0.66 MeV with maximum penetration at same dose [6, 9]. Study suggests no change in allergenicity upon 25 kGy irradiation dose of ^{60}Co γ- exposure on raw kidney bean or black gram seeds [102], and on soybean seeds upon multiple exposure ranging from 2.5 to 30 kGy radiation [103]. Same group also found that co-administration of γ-irradiation (25 kGy) and boiling at 121°C for 15 minutes remarkably suppressed the immunogenicity of soluble and insoluble allergens in kidney bean and black gram seeds against untreated control condition [102]. Another study suggests that γ-irradiation <= 5 kGy administration to soybean protein solution exposed the conformational and linear epitopes of allergen proteins, thus, augments its immunogenecity and γ-irradiation > = 10 kGy suppressed the immunogenicity by 91% compared to control by inducing the protein-degradation or aggregation cellular event masking the conformational epitopes responsible for allergenicity [27, 28]. Collectively, the immonosensitivity of soybean allergen proteins depends on the state of its conformational changes induced by γ-irradiation and additional processing factors; however, its clinical execution warrants further in-depth research.

3. Effects of γ-irradiation in reducing microbial contamination load in soybean

Soybean stands as one of the most vital grains globally, yet fungal infections are an inevitable challenge in its cultivation and storage. Pathogens like Fusarium, Alternaria, and Phomopsis inflict substantial damage on the quality and market value of soybeans [104]. Because of the existence of fungi that produce toxins, soybeans afflicted by fungal damage are unfit for human consumption. Interestingly, these fungi-damaged soybeans boast higher protein content and a comparable amino acid composition to their uninfected counterparts [105]. Consequently, certain portions of these fungi-affected soybeans have been repurposed as raw materials for animal feed by blending them with healthy soybeans [104]. Nonetheless, they quite pose risks to animals. Hence, the elimination of fungi and mycotoxins becomes of paramount importance. γ-irradiation has emerged as a secure and efficient physical approach for decontaminating microbes (**Figure 1**), controlling pests, and enhancing the quality of both unprocessed and processed agricultural products [106]. Research has provided evidence that γ-irradiation at moderate doses, typically ranging from 8 to 10 kGy, can completely eliminate bacteria and fungi from oilseeds while preserving their chemical composition [59, 107]. Moreover, high-dose γ-irradiation (10–30 kGy) has proven effective in reducing mycotoxins, including T-2 toxin, aflatoxins, zearalenone, ochratoxin, and deoxynivalenol, produced by specific fungal strains in food crops [108, 109]. It's important to note that higher irradiation doses lead to a more comprehensive degradation of mycotoxins. Moreover, γ-irradiation has demonstrated its effectiveness in decreasing the ruminal degradability of soybean protein and enhancing in vitro crude protein digestibility [69].

γ-irradiation has demonstrated its effectiveness for controlling fungal growth and deactivating mycotoxins, which are highly toxic substances. Given the severe health risks associated with mycotoxins, various methods have been employed to eliminate their presence in various food products. The application of ionizing radiation treatment has been well-established as an effective physical method for preserving the quality of food and prolonging the shelf life of agricultural products, including soybeans [110, 111]. Aflatoxins are renowned secondary metabolites generated by fungi including *Aspergillus nomius*, *Aspergillus parasiticus*, and *Aspergillus flavus* [112]. The sources of these fungi have been linked to groundnut meal contaminated with aflatoxin-producing fungi, which is commonly used in animal feed [113]. Given their virulence, carcinogenicity, mutagenicity, and teratogenicity, AFs are regarded as highly hazardous substances that pose risks to animal and human health [114, 115]. Among the 20 different types of aflatoxins, including AFB1, AFB2, AFG1, and AFG2, are frequently detected in various food products [116].

AFs exhibit remarkable resistance to heat treatment, as their decomposition temperature exceeds 235°C [117–119]. As a result, conventional drying methods alone do not lead to a substantial reduction in their concentrations in stored grains. However, prolonged exposure to elevated temperatures seems to have a beneficial impact on decontamination. For instance, subjecting soybeans to heat treatments at 100°C and 150°C for 90 minutes resulted in notable reductions of AFB1 contents by 41.9% and 81.2%, respectively [120]. γ-irradiation has proven to be effective in decreasing both fungal contamination and AFB1 concentration in naturally infected corn kernels. γ-irradiation doses in the range of 1 to 10 kGy have resulted in mycotoxin level reductions ranging from 69.8% to 94.5%, respectively [121, 122]. Additionally, for wheat, rice and corn kernels, doses of 4, 6, and 8 kGy have proven effective in reducing AF levels by 15–56% with increasing doses [123]. In the case of soybeans, doses exceeding 10 kGy have shown utility in AFB1 reduction [124]. Ozone treatment has shown remarkable efficiency in the destruction of AFs, achieving reductions of up to 66–95% of the original toxin levels in cereal flours, grains, soybeans, and peanuts [125–127]. While Hooshmand and Klopenstein [108] found that γ-irradiation doses up to 20 kGy did not have a notable impact on the AFB1 content in soybeans, corn, and wheat. γ-irradiation reduced the yeast and mold populations in soybeans, with yeast and mold becoming undetectable at doses exceeding 5.0 kGy. The application of a 3.0 kGy dose of γ-irradiation effectively lowered the aerobic bacterial populations in soybeans to a level that met acceptable standards from 5.77×10^5 to 1.2×10^2 CFU/g [51].

Furthermore, in the study by Wilson et al. [104], a positive correlation was noted between the fat content of soybeans and fungal damage, which they attributed to a decrease in seed mass. Dogbevi et al. [128] demonstrated that γ-irradiation at 1.5 kGy led to a reduction in the mold population in dry red kidney beans by a factor of 2 log cycles. Additionally, γ-irradiation at doses of 1.0, 3.0, 5.0, and 10.0 kGy resulted in significantly lower counts of total aerobic bacteria in soybeans compared to the control. The counts of mold and yeast in soybeans were also reduced by irradiation, with no presence of yeast or mold in soybeans exposed to doses of 5.0 and 10.0 kGy.

4. Risks associated with high dose γ-irradiation on soybean nutrients

So far, we have unraveled numerous beneficial impacts of recommended low dose γ-irradiation (0.5 to 10 kGy) in reducing the soybean microbial contamination, allergenicity, ANFs as well as increasing nutrient composition, flavor, shelf-life, anti-oxidant, and ROS scavenging activities. Impact of various doses of y-irradiation also depends on the soybean state (seed, flour, protein solution, flavor, etc.) and its intended use. Numerous studies suggest a high dose of γ-irradiation (>10 to 400 kGy) exerts significant deleterious effects on soybean nutrients, protein conformation, and various intrinsic properties. Increased γ-irradiation using ^{60}Co induce changes in the secondary structure of protein and exposes hydrophobic amino acids, thus, increasing protein insolubility and a simultaneous decrease in its emulsifying ability and reduced vitamin-E level was observed [21, 22]. A similar dose further decreases α-tocopherol (an anti-oxidant) by 35.4% compared to the control, increases ROS (hydroperoxide and peroxide) generation [56], as well as proanthocyanidins [46]. Very high radiation dose (400 to 1200 kGy of γ-irradiation) significantly induces the soluble dietary fiber constituent in soybean [34]. In continuation, a dose-dependent increase in soybean trypsin inhibitor activity has been demonstrated by Farag [62]. While γ-radiation can deactivate trypsin inhibitor (TI), an antinutrient and food allergen in soybeans, but its effectiveness depends on moisture levels. In an aqueous solution, TI is almost completely destroyed at 10 kGy irradiation of Cobalt-60. However, in a dry state or with 50% moisture, significant functional and structural losses occurs at higher doses (100 kGy and 30 kGy, respectively). Surprisingly, TI remains unaffected in other legumes, as well as in soaked and dry soybean seeds, even at substantial irradiation doses (50 kGy and 100 kGy respectively). Hence, TI is stable to direct gamma radiation, and its inactivation at lower doses requires excess moisture. Using gamma radiation to reduce TI in dry or soaked seeds would necessitate high doses that could impact its sensory and functional propertie [129]. Interestingly, the impact of y-irradiation on soybean allergenicity is diverse as demonstrated by increased allergenicity at <5 kGy [102] and decreasing allergenicity at >10 kGy [27, 28] by unmasking and masking of linear epitopes of allergen proteins respectively. Increased γ-irradiation (>10 to 30 kGy) dose is also associated with reduced fat, protein, lipase, and anti-oxidant enzyme activities in soybean as reported in previous sections.

5. Conclusion

Irradiation, a cost-effective and environmentally friendly non-thermal food processing technology, is widely employed in the food industry. It has been found to have various benefits, such as reducing food allergens and effectively eliminating insects, molds, and bacteria. Additionally, it aids in reducing the presence of detrimental compounds such as biogenic amines and substances that hinder nutrient absorption in food. When used at appropriate doses, γ-irradiation does not adversely affect the physical, chemical, nutritional, or sensory qualities of food products. For instance, γ-irradiation doses of up to 10 kGy have been shown to have a minimal adverse impact on the natural physicochemical properties, nutritional value, and flavor of soybean oil. Studies indicate that a gamma irradiation dose of 10.0 kGy is a successful method for inhibiting microbes' infestation and decontaminating soybean seeds. Moreover, γ-irradiation at these doses increases isoflavone content while reduces tocopherol contents and the compounds responsible for flatulence (RFOs). At moderate radiation doses ranging from 8 to 10 kGy, gamma irradiation can entirely eradicate bacteria and fungi in oilseeds while leaving their composition unaffected. Assessments based on sensory criteria have shown that soybeans subjected to irradiation remain satisfactory even at doses as high as 5.0 kGy. Low-dose γ-irradiation appears to enhance the total phenolic content and increase proanthocyanidins, contributing to improved anti-oxidant activity. Furthermore, high-dose γ-irradiation (10–30 kGy) has been shown to reduce mycotoxins, including aflatoxins and other toxins produced by fungal infections.

γ-irradiation has the added benefit of enhancing the nutritional quality of soybeans by reducing anti-nutritional components, such as phytic acid, trypsin inhibitors, lipoxygenase, oligosaccharides, tannins, urease 3, and hemagglutinating agents. It's important to highlight that when it comes to reducing the allergenic properties of soybeans, using gamma irradiation at doses lower than 10 kGy may not achieve the intended outcomes. On the other hand, it has been demonstrated that employing higher doses exceeding 10 kGy can effectively diminish the allergenic properties of soybean protein solutions. High doses of γ-irradiation (> 10 kGy) significantly impact soybean nutrients (reduced fat and protein content) and intrinsic properties (like reduced anti-oxidant, lipase activity, increased UFA, ANFs, etc.). In summary, γ-irradiation is a valuable tool in the soybean food industry, offering multiple benefits such as microbial decontamination, enhanced shelf-life, nutritional quality, flavor, and improved safety, with effectiveness and outcomes depending on the specific dose, duration, and other cotreatments (like temperature, pressure, micropulverization, enzyme catalysis, etc.) applied.

Conflict of interest

The authors declare no conflict of interest.

Author details

Kalpana Tewari[1†], Mahipal Singh Kesawat[2†], Vinod Kumar[3], Chirag Maheshwari[4], Veda Krishnan[4], Sneh Narwal[4], Sweta Kumari[4*], Anil Dahuja[4*], Santosh Kumar[5*] and Swati Manohar[6*]

1 Department of Basic Science, ICAR-Indian Institute of Pulse Research (IIPR), Kanpur, Uttar Pradesh, India

2 Faculty of Agriculture, Department of Genetics and Plant Breeding, Sri Sri University, Cuttack, Odisha, India

3 Department of Molecular Biology and Genetic Engineering, Bihar Agricultural University, Bhagalpur, Bihar, India

4 Division of Biochemistry, ICAR-Indian Agricultural Research Institute (IARI), New Delhi, India

5 Department of Bioscience and Bioengineering, Indian Institute of Technology, Kanpur, Uttar Pradesh, India

6 Department of Biotechnology, School of Life Sciences, Mahatma Gandhi Central University, East Champaran, Bihar, India

†Equally contributed.

*Address all correspondence to: sweta381@gmail.com, ad_bio@yahoo.com, santoshszu1517@outlook.com and swatibiotech@mgcub.ac.in

References

[1] Sinanoglou VJ, Batrinou A, Konteles S, Sflomos K. Microbial population, physicochemical quality, and allergenicity of molluscs and shrimp treated with cobalt-60 gamma radiation. Journal of Food Protection. 2007;**70**(4):958-966

[2] Zoumpoulakis P, Sinanoglou VJ, Batrinou A, Strati IF, Miniadis-Meimaroglou S, Sflomos K. A combined methodology to detect γ-irradiated white sesame seeds and evaluate the effects on fat content, physicochemical properties and protein allergenicity. Food Chemistry. 2012;**131**(2):713-721

[3] Křížek M, Matějková K, Vácha F, Dadáková E. Effect of low-dose irradiation on biogenic amines formation in vacuum-packed trout flesh (Oncorhynchus mykiss). Food Chemistry. 2012;**132**(1):367-372

[4] Mallikarjunan N, Marathe S, Rajalakshmi D, Mahesh S, Jamdar SN, Sharma A. Effect of ionizing radiation on structural and functional attributes of red kidney bean (*Phaseolus vulgaris* L.) lectin. LWT-Food Science and Technology. 2014;**59**(1):300-307

[5] An BJ, Kwak JH, Son JH, Park JM, Lee JY, Jo C, et al. Biological and anti-microbial activity of irradiated green tea polyphenols. Food Chemistry. 2004;**88**(4):549-555

[6] High-dose irradiation: Wholesomeness of food irradiated with doses above 10 kGy. Report of a Joint FAO/IAEA/WHO Study Group. World Health Organization Technical Report Series. 1999:**890**:i-vi:1-197

[7] Robichaud V, Bagheri L, Aguilar-Uscanga BR, Millette M, Lacroix M. Effect of ɣ-irradiation on the microbial inactivation, nutritional value, and antioxidant activities of infant formula. Lebensmittel-Wissenschaft & Technologie. 2020;**125**:109211

[8] Liu K, Liu Y, Chen F. Effect of gamma irradiation on the physicochemical properties and nutrient contents of peanut. Lebensmittel-Wissenschaft & Technologie. 2018;**96**:535-542. DOI: 10.1016/j.lwt.2018.06.009

[9] Fan L, Han P, Feng X, Duan X, Li M, Zhang B, et al. Research progress in the application development of electron beam irradiation in food industry. Science and Technology of Food Industry. 2013;**14**:374-380. DOI: 10.13386/j.issn1002-0306.2014.14.074

[10] Chizoba, Ekezie FG, Cheng JH, Sun DW. Effects of nonthermal food processing technologies on food allergens: A review of recent research advances. Trends in Food Science & Technology. 2018;**74**:12-25. DOI: 10.1016/j.tifs.2018.01.007

[11] Oliveira CL, Hoz LDL, Silva JC, Torriani IL, Netto FM. Effects of gamma radiation on β-lactoglobulin: Oligomerization and aggregation. Biopolymers: Original Research on Biomolecules. 2007;**85**(3):284-294

[12] McParland BJ, McParland BJ. Biological effects of ionizing radiation. In: Nuclear Medicine Radiation Dosimetry: Advanced Theoretical Principles. 2010. pp. 401-453

[13] Abu JO, Müller K, Duodu KG, Minnaar A. Gamma irradiation of cowpea (Vigna unguiculata L. Walp) flours and pastes: Effects on functional,

thermal and molecular properties of isolated proteins. Food Chemistry. 2006;**95**(1):138-147

[14] Zhao Y, Sun N, Li Y, Cheng S, Jiang C, Lin S. Effects of electron beam irradiation (EBI) on structure characteristics and thermal properties of walnut protein flour. Food Research International. 2017;**100**:850-857

[15] General Standard for Irradiated Foods. CODEX STAN 106-1983, Rev.1-2003. pp. 250-254

[16] Steele JH. Food irradiation: A public health challenge for the 21st century. Clinical Infectious Diseases. 2001;**33**(3):376-377. DOI: 10.1086/321899

[17] Farkas J, Mohácsi-Farkas C. History and future of food irradiation. Trends in Food Science & Technology. 2011;**22**(2-3):0-126. DOI: 10.1016/j.tifs.2010.04.002

[18] Byun MW, Kang IJ, Kwon JH, Hayashi Y, Mori T. Physicochemical properties of soybean oil extracted from γ-irradiated soybeans. Radiation Physics and Chemistry. 1996;**47**(2):301-304. DOI: 10.1016/0969-806x(94)00149-e

[19] Saadet KG, Zeki BS, Hilmi ÇA. Effects of gamma irradiation on chemical and sensory characteristics of natural hazelnut kernels. Postharvest Biology and Technology. 2017;**123**:12-21. DOI: 10.1016/j.postharvbio.2016.08.007

[20] Zhang ZS, Li CY, Liu YL. Effect of mildew and γ-irradiation on the physicochemical property and structure of soybean proteins. Modern Food Science and Technology. 2015;**31**(7):191-196. DOI: 10.13982/j.mfst.1673-9078.2015.7.030

[21] Wang Y, Zhang A, Wang Y, Wang X, Xu N, Jiang L. Effects of irradiation on the structure and properties of glycosylated soybean proteins. Food & Function. 2020a;**11**(2):1635-1646. DOI: 10.1039/c9fo01879d

[22] Wang Y, Guldiken B, Tulbek M, House JD, Nickerson M. Impact of alcohol washing on the flavour profiles, functionality and protein quality of air-classified pea protein enriched flour. Food Research International. 2020b;**132**:109085

[23] Shi Y, Li Ru-yi, Tu Zong-cai, Ma Da, Wang H, Huang Xiao-qin, He Na. Effect of γ-irradiation on the physicochemical properties and structure of fish myofibrillar proteins. Radiation Physics and Chemistry. 2015; 109: 70-72. doi:10.1016/j.radphyschem.2014.12.016.

[24] Afify AMR, Rashed MM, Ebtesam AM, Elbeltagi HS. Effect of gamma radiation on the lipid profiles of soybean, peanut and sesame seed oils. Grasas Y Aceites. 2013;**64**(4):356-368

[25] Kumar V, Rani A, Dixit AK, Pratap D, Bhatnagar D. A comparative assessment of total phenolic content, ferric reducing-anti-oxidative power, free radical-scavenging activity, vitamin C and isoflavones content in soybean with varying seed coat colour. Food Research International. 2010;**43**(1):323-328

[26] Messina M. Impact of soy foods on the development of breast cancer and the prognosis of breast cancer patients. Complementary Medicine Research. 2016;**23**(2):75-80

[27] Meinlschmidt P, Ueberham E, Lehmann J, Reineke K, Schl€uter O, Schweiggert-Weisz U, et al. The effects of pulsed ultraviolet light, cold atmospheric pressure plasma, and gamma irradiation on the immunoreactivity of soy protein isolate. Innovative Food Science & Emerging Technologies. 2016a;**38**:374-383. DOI: 10.1016/j.ifset.2016.06.007

[28] Meinlschmidt P, Sussmann D, Schweiggert-Weisz U, Eisner P. Enzymatic treatment of soy protein isolates: Effects on the potential allergenicity, technofunctionality, and sensory properties. Food Science & Nutrition. 2016b;**4**(1):11-23

[29] Wang B, Zhang Q, Zhang N, Bak KH, Soladoye OP, Aluko RE, et al. Insights into formation, detection and removal of the beany flavor in soybean protein. Trends in Food Science & Technology. 2021;**112**:336-347

[30] Roland WS, Pouvreau L, Curran J, van de Velde F, de Kok PM. Flavor aspects of pulse ingredients. Cereal Chemistry. 2017;**94**(1):58-65

[31] Yu H, Liu R, Hu Y, Xu B. Flavor profiles of soymilk processed with four different processing technologies and 26 soybean cultivars grown in China. International Journal of Food Properties. 2017;**20**(sup3):S2887-S2898

[32] deMan JM, Finley JW, Hurst WJ, Lee CY. Principles of Food Chemistry. Fourth ed. Cham: Food Science Text Series. Springer International Publishing; 2018

[33] Cheng T, Liu C, Hu Z, Wang Z, Guo Z. Effects of γ-irradiation on structure and functional properties of pea Fiber. Food. 2022;**11**(10):1433. DOI: 10.3390/foods11101433

[34] Zhu L, Yu B, Chen H, Yu J, Yan H, Luo Y, et al. Comparisons of the micronization, steam explosion, and gamma irradiation treatment on chemical composition, structure, physicochemical properties, and in vitro digestibility of dietary fiber from soybean hulls. Food Chemistry. 2022;**366**:130618. DOI: 10.1016/j.foodchem.2021.130618

[35] Tewari K, Kumari S, Vinutha T, Singh B, Dahuja A. Gamma irradiation induces reduction in the off-flavour generation in soybean through enhancement of its antioxidant potential. Journal of Radioanalytical and Nuclear Chemistry. 2015;**303**:2041-2051

[36] Kumari S, Gupta OP, Mishra CB, Thimmegowda V, Krishnan V, Singh B, et al. Gamma irradiation, an effective strategy to control the oxidative damage of soy proteins during storage and processing. Radiation Physics and Chemistry. 2020;**177**:109134

[37] Farkhad SA, Hosseini A. Effect of gamma irradiation on antioxidant potential, isoflavone aglycone and phytochemical content of soybean (Glycine max L. Merrill) cultivar Williams. Journal of Radioanalytical and Nuclear Chemistry. 2020;**324**:497-505. DOI: 10.1007/s10967-020-07100-0

[38] de Camargo AC, de Souza VTM, Regitano-D'Arce MA, Calori-Domingues MA, Canniatti-Brazaca SG. Gamma radiation effects on peanut skin antioxidants. International Journal of Molecular Sciences. 2012;**13**(3):3073-3084

[39] Luo X, Du Z, Yang K, Wang J, Zhou J, Liu J, et al. Effect of electron beam irradiation on phytochemical composition, lipase activity and fatty acid of quinoa. Journal of Cereal Science. 2021;**98**:103161

[40] Popović BM, Stajner D, Mandić A, Canadanović-Brunet J, Kevrešan S. Enhancement of antioxidant and isoflavones concentration in gamma irradiated soybean. Scientific World Journal. 2013;**2013**:383574. DOI: 10.1155/2013/383574

[41] Variyar PS, Limaye A, Sharma A. Radiation-induced enhancement of antioxidant contents of soybean (Glycine max Merrill). Journal of Agricultural and

Food Chemistry. 2004;**52**(11):3385-3388. DOI: 10.1021/jf030793j

[42] Aguiar CL, Baptista AS, Walder JM, Tsai SM, Carrão-Panizzi MC, Kitajima EW. Changes in isoflavone profiles of soybean treated with gamma irradiation. International Journal of Food Sciences and Nutrition. 2009;**60**(5):387-394. DOI: 10.1080/09637480701754968

[43] Krishnan V, Gothwal S, Dahuja A, Vinutha T, Singh B, Jolly M, et al. Enhanced nutraceutical potential of gamma irradiated black soybean extracts. Food Chemistry. 2018;**245**:246-253

[44] Krishnan V, Singh A, Thimmegowda V, Singh B, Dahuja A, Rai RD, et al. Low gamma irradiation effects on protein profile, solubility, oxidation, scavenger ability and bioavailability of essential minerals in black and yellow Indian soybean (*Glycine max* L.) varieties. Journal of Radioanalytical and Nuclear Chemistry. 2016;**307**:49-57

[45] Bansal N, Dahiya P, Prashat GR, Goswami D, Kumari S, Pushkar S, et al. Effects of gamma irradiation on soybean oil stability by enhancing tocopherol content in soybean. Journal of Radioanalytical and Nuclear Chemistry. 2020;**326**(3):1617-1629

[46] Štajner D, Milošević M, Popović BM. Irradiation effects on phenolic content, lipid and protein oxidation and scavenger ability of soybean seeds. International Journal of Molecular Sciences. 2007;**8**(7):618-627

[47] Mata-Ramírez D, Serna-Saldívar S, Antunes-Ricardo M. Enhancement of anti- inflammatory and antioxidant metabolites in soybean (Glycine max) calluses subjected to selenium or UV-light stresses. Scientia Horticulturae Amsterdam. 2019;**257**:108669. DOI: 10.1016/j.scienta.2019

[48] Wang M, Liu G, Guo T, Xie C, Wang P, Yang R. UV-B radiation enhances isoflavone accumulation and antioxidant capacity of soybean calluses. Frontiers in Nutrition. 2023;**10**:1139698. DOI: 10.3389/fnut.2023.1139698

[49] Dixit AK, Bhatnagar D, Kumar V, Rani A, Manjaya JG, Bhatnagar D. Gamma irradiation induced enhancement in isoflavones, total phenol, anthocyanin and antioxidant properties of varying seed coat colored soybean. Journal of Agricultural and Food Chemistry. 2010;**58**(7):4298-4302

[50] Braunrath R, Isnardy B, Solar S, Elmadfa I. Influence of α-, γ-, and δ-tocopherol on the radiation induced formation of peroxides in rapeseed oil triacylglycerols. Food Chemistry. 2009;**117**:349-351

[51] Yun J, Li X, Fan X, Tang Y, Xiao Y, Wan S. Effect of gamma irradiation on microbial load, physicochemical and sensory characteristics of soybeans (*Glycine max* L. Merrill). Radiation Physics and Chemistry. 2012;**81**:1198-1202

[52] Dixit AK, Kumar V, Rani A, Manjaya JG, Bhatnagar D. Effect of gamma irradiation on lipoxygenases, trypsin inhibitor, raffinose family oligosaccharides and nutritional factors of different seed coat colored soybean (*Glycine max* L.). Radiation Physics and Chemistry. 2011;**80**:597-603

[53] Minami I, Nakamura Y, Todoriki S, Murata Y. Effect of γ irradiation on the fatty acid composition of soybean and soybean oil. Bioscience, Biotechnology, and Biochemistry. 2012;**76**(5):900-905. DOI: 10.1271/bbb.110859

[54] Hafez YS, Mohamed AI, Singh G, Hewedy FM. Effects of gamma irradiation on proteins and fatty acids of soybean. Journal of Food Science. 1985;**50**:1271-1274

[55] Aziz NH, Souzan RM, Shahin AA. Effect of γ-irradiation on the occurrence of pathogenic microorganisms and nutritive value of four principal cereal grains. Applied Radiation and Isotopes. 2006;**64**(12):1555-1562. DOI: 10.1016/j.apradiso.2005.10.006

[56] Mahrous SR. Chemical properties of aspergillus flavus-infected soybean seeds exposed to γ-irradiation during storage. International Journal of Agriculture And Biology. 2007;**9**:231-238

[57] Mexis SF, Kontominas MG. Effect of gamma irradiation on the physico-chemical and sensory properties of raw shelled peanuts (*Arachis hypogaea* L.) and pistachio nuts (*Pistacia vera* L.). Journal of the Science of Food and Agriculture. 2009;**89**:867-875. DOI: 10.1002/jsfa.3526

[58] Bhatti IA, Iqbal M, Anwar F, Shahid SA, Shahid M. Quality characteristics and microbiological safety evaluation of oils extracted from gamma irradiated almond (Prunus dulcis mill.) seeds. Grasas y Aceites. 2013;**64**:68-76

[59] Bhatti IA, Ashraf S, Shahid M, Asi MR, Mehboob S. Quality index of oils extracted from γ-irradiated peanuts (*Arachis hypogaea* L.) of the golden and bari varieties. Applied Radiation and Isotopes. 2010;**68**(12):2197-2201. DOI: 10.1016/j.apradiso.2010.05.017

[60] Vasconcelos IM, Maia AAB, Siebra EA, Oliveira JTA, de FFU CA, VMM M, et al. Nutritional study of two Brazilian soybean (*Glycine max*) cultivars differing in the contents of antinutritional and toxic proteins. The Journal of Nutritional Biochemistry. 2001;**12**(1):55-62. DOI: 10.1016/s0955-2863(00)00148-0

[61] Abu-Tarboush HM. Irradiation inactivation of some antinutritional factors in plant seeds. Journal of Agricultural and Food Chemistry. 1998;**46**:2698-2702

[62] Farag MD, El-Din H. The nutritive value for chicks of full-fat soybeans irradiated at up to 60kGy. Animal Feed Science and Technology. 1998;**73**(3-4):319-328. DOI: 10.1016/s0377-8401(98)00148-5

[63] Villavicencio ALCH, Mancini-Filho J, Delincee H, Greiner R. Effect of irradiation on anti-nutrients (total phenolics, tannins and phytate) in Brazilian beans. Radiation Physics and Chemistry. 2000;**57**(3-6):289-293. DOI: 10.1016/s0969-806x(99)00393-x

[64] Mechi R, Canniatti-Brazaca SG, Arthur V. Avaliac-a˜o quı′mica, nutricional e fatores antinutricionais do feija˜o preto (Phaseolus vulgaris L.) irradiado. Food Science and Technology. 2005;**25**:109-114

[65] Amaya H, Acevedo E, Bressani R. Efecto del recalentamiento sobre la disponibilidad de hierro y el valor nutritivo de la proteı′na del frijol negro (*Phaseolus vulgaris*) cocido. Archivos Latinoamericanos de Nutrición. 1991;**41**:223-237

[66] Brigide P, Canniatti-Brazaca SG. Antinutrients and in vitro availability iron of irradiated common beans (Phaseolus vulgaris). Food Chemistry. 2006;**98**:85-89

[67] Pino VHD, Lajolo FM. Efecto inhibitorio de los taninos del frijol carioca (Phaseolus vulgaris) sobre la digestibilidad de la faseolina por dos

sistemas multienzimaticos. Food Science and Technology. 2003;**23**:49-53

[68] De-Toledo TCF, Canniatti-Brazaca SG, Arthur V, Piedade SMS. Effects of gamma radiation on total phenolics, trypsin and tannin inhibitors in soybean grains. Radiation Physics and Chemistry. 2007;**76**:1653-1656

[69] Taghinejad M, Nikkhah A, Sadeghi AA, Raisali G, Chamani M. Effects of gamma irradiation on chemical composition, antinutritional factors, ruminal degradation and in vitro protein digestibility of full-fat soybean. Asian-Australasian Journal of Animal Sciences. 2009;**22**:534-541

[70] Khalil A, Al-Bachir M. Decontamination of polycyclic aromatic hydrocarbons in pea seeds by gamma irradiation. Journal of Food Measurement and Characterization. 2017;**1**:1-7

[71] Lee KJ, Hwang JE, Velusamy V, Ha BK, Kim JB, Kim SH, et al. Selection and molecular characterization of a lipoxygenase-free soybean mutant line induced by gamma irradiation. Theoretical and Applied Genetics. 2014;**127**(11):2405-2413. DOI: 10.1007/s00122-014-2385-9

[72] Yuan FJ, Zhu DH, Tan YY, Dong DK, Fu XJ, Zhu SL, et al. Identification and characterization of the soybean IPK1 ortholog of a low phytic acid mutant reveals an exon-excluding splice-site mutation. Theoretical and Applied Genetics. 2012;**125**:1413-1423

[73] Hayashi M, Pawankar R, Yamanishi S, Itoh Y. Food-dependent exercise-induced anaphylaxis to soybean: Gly m 5 and Gly m 6 as causative allergen components. World Allergy Organization Journal. 2020;**13**(7):100439

[74] Xia J, Zu Q, Yang A, Wu Z, Li X, Tong P, et al. Allergenicity reduction and rheology property of lactobacillus-fermented soymilk. Journal of the Science of Food and Agriculture. 2019;**99**(15):6841-6849

[75] Riascos JJ, Weissinger SM, Weissinger AK, Kulis M, Burks AW, Pons L. The seed biotinylated protein of soybean (Glycine max): A boiling-resistant new allergen (Gly m 7) with the capacity to induce IgE-mediated allergic responses. Journal of Agricultural and Food Chemistry. 2016;**64**(19):3890-3900

[76] Wang T, Qin GX, Sun ZW, Zhao Y. Advances of research on glycinin and β-conglycinin: A review of two major soybean allergenic proteins. Critical Reviews in Food Science and Nutrition. 2014;**54**(7):850-862

[77] Frias J, Song YS, Martínez-Villaluenga C, De Mejia EG, Vidal-Valverde C. Immunoreactivity and amino acid content of fermented soybean products. Journal of Agricultural and Food Chemistry. 2008;**56**(1):99-105

[78] Holzhauser T, Wackermann O, Ballmer-Weber BK, Bindslev-Jensen C, Scibilia J, Perono-Garoffo L, et al. Soybean (Glycine max) allergy in Europe: Gly m 5 (β-conglycinin) and Gly m 6 (glycinin) are potential diagnostic markers for severe allergic reactions to soy. Journal of Allergy and Clinical Immunology. 2009;**123**:452-458

[79] Ballmer-Weber BK, Holzhauser T, Scibilia J, Mittag D, Zisa G, Ortolani C, et al. Clinical characteristics of soybean allergy in Europe: A double-blind, placebo-controlled food challenge study. The Journal of Allergy and Clinical Immunology. 2007;**119**:1489-1496

[80] Becker W, Brasseur D, Bresson JL, Flynn A, Jackson AA, Lagiou P, et al.

Opinion of the scientific panel on dietetic products, nutrition and allergies on a request from the commission relating to the evaluation of allergenic foods for labelling purposes (request nr EFSA-Q-2003-016). EFSA Journal. 2004;**32**:1-197

[81] Heimo B, Mills ENC. Plant food allergens-structural and functional aspects of allergenicity. Biotechnology Advances. 2005;**23**:395-399

[82] Sun P, Li DF, Li ZJ, Dong B, Wang FL. Effects of glycinin on IgE-mediated increase of mast cell numbers and histamine release in the small intestine. Journal of Nutritional Biochemistry. 2008;**19**(9):627-633

[83] Adachi A, Horikawa T, Shimizu H, Sarayama Y, Ogawa T, Sjolander S. Soybean β-conglycinin as the main allergen in a patient with food-dependent exercise-induced anaphylaxis by tofu: Food processing alters pepsin resistance. Clinical and Experimental Allergy. 2008;**39**:167-173

[84] Sub-Committee WIA N. 2021. Available from: http://www.allergen.org/index.php

[85] Tsai JJ, Chang CY, Liao EC. Comparison of allergenicity at Gly m 4 and Gly m bd 30K of soybean after genetic modification. Journal of Agricultural and Food Chemistry. 2017;**65**(6):1255-1262

[86] Ogawa T, Bando N, Tsuji H, Nishikawa K, Kitamura K. Bioscience, Biotechnology, and Biochemistry. 1995;**59**:831-833

[87] Moroz LA, Yang WH. The New England Journal of Medicine. 1980;**302**:1126-1128

[88] Thom de Souza CC, Rosario Filho NA, Camargo JF, Godoi RHM. Levels of airborne soybean allergen (Gly m 1) in a Brazilian soybean production city: A pilot study. International Journal of Environmental Research and Public Health. 2020;**17**(15):5381

[89] Gonzalez RDO, Calabozo BBD, Carreira JPF. Monoclonal antibody-based method to quantify Gly m 1. Its application to assess environmental exposure to soybean dust. Allergy. 2000;**55**:59-64

[90] Amnuaycheewa P, de Mejia EG. Purification, characterisation, and quantification of the soy allergen profilin (Gly m 3) in soy products. Food Chemistry. 2010;**119**(4):1671-1680

[91] Holzhauser T, Franke A, Treudler R, Schmiedeknecht A, Randow S, Becker WM, et al. The BASALIT multicenter trial: Gly m 4 quantification for consistency control of challenge meal batches and toward Gly m 4 threshold data. Molecular Nutrition & Food Research. 2017;**61**(3):1600527

[92] Hanafusa K, Murakami H, Ueda T, Yano E, Zaima N, Moriyama T. Worm wounding increases levels of pollen-related food allergens in soybean (*Glycine max*). Bioscience Biotechnology & Biochemistry. 2018;**82**(7):1207-1215

[93] Mo X, Wang D, Sun XS. Physicochemical properties of beta and alpha'alpha subunits isolated from soybean beta-conglycinin. Journal of Agricultural and Food Chemistry. 2011;**59**(4):1217-1222

[94] Lacorn M, Dubois T, Siebeneicher S, Weiss T. Accurate and sensitive quantification of soy proteins in raw and processed food by sandwich ELISA. Food Science and Technology. 2016;**4**(4):69-77

[95] Geng T, StojsIn DK, Liu K, Schaalje B, Postin C, Ward J, et al. Natural variability

of allergen levels in conventional soybeans: Assessing variation across north and South America from five production years. Journal of Agricultural and Food Chemistry. 2017;**65**(2):463-472

[96] Xiang P, Haas EJ, Zeece MG, Markwell J, Sarath G. C-terminal 23 kDa polypeptide of soybean Gly m bd 28 K is a potential allergen. Planta. 2004;**220**(1):56-63

[97] Pi X, Yang Y, Sun Y, Cui Q, Wan Y, Fu G, et al. Recent advances in alleviating food allergenicity through fermentation. Critical Reviews in Food Science and Nutrition. 2021;**1**:1-14

[98] Berneder M, Bublin M, Hoffmann-Sommergruber K, Hawranek T, Lang R. Allergen chip diagnosis for soy-allergic patients: Gly m 4 as a marker for severe food-allergic reactions to soy. International Archives of Allergy and Immunology. 2013;**161**(3):229-233

[99] Yang A, Zuo L, Cheng Y, Wu Z, Li X, Tong P, et al. Degradation of major allergens and allergenicity reduction of soybean meal through solid-state fermentation with microorganisms. Food & Function. 2018;**9**(3):1899-1909

[100] Fukutomi Y, Sjolander S, Nakazawa T, Borres MP, Ishii T, Nakayama S, et al. Clinical relevance of IgE to recombinant Gly m 4 in the diagnosis of adult soybean allergy. The Journal of Allergy and Clinical Immunology. 2012;**129**(3):860-863 e863

[101] Farkas J. Irradiation for better foods. Trends in Food Science & Technology. 2006;**17**:148-152

[102] Kasera RAB, Singh R, Kumar S, Lavasa KN, Prasad AN. Effect of thermal processing and gamma-irradiation on allergenicity of legume proteins. Food and Chemical Toxicology. 2012;**50**(10):3456-3461. DOI: 10.1016/j.fct.2012.07.031

[103] Moriyama T, Yano E, Kitta K, Kawamoto S, Kawamura Y, Todoriki S. Effect of gamma irradiation on soybean allergen levels. Bioscience, Biotechnology, and Biochemistry. 2013;**77**(12):2371-2377. DOI: 10.1271/bbb.130487

[104] Wilson RF, Novitzky WP, Fenner GP. Effect of fungal damage on seed composition and quality of soybeans. Journal of the American Oil Chemists' Society. 1995;**72**:1425-1429

[105] Meriles JM, Lamarque AL, Labuckas DO, Maestri DM. Effect of fungal damage by fusarium spp and diaporthe/phomopsis complex on protein quantity and quality of soybean seed. Journal of the Science of Food & Agriculture. 2004;**84**(12):1594-1598

[106] Calado T, Venâncio A, Abrunhosa L. Irradiation for mold and mycotoxin control: A review. Comprehensive Reviews in Food Science and Food Safety. 2014;**13**:1049-1061

[107] Arici M, Colak FA, Gecgel U. Effect of gamma radiation on microbiological and oil properties of black cumin (Nigella sativa L.). Grasas y Aceites. 2007;**58**:339-343

[108] Hooshmand H, Klopfenstein CF. Effects of gamma irradiation on mycotoxin disappearance and amino acid contents of corn, wheat, and soybeans with different moisture contents. Plant Foods for Human Nutrition. 1995;**47**(3):227-238. DOI: 10.1007/BF01088331

[109] Jalili M, Jinap S, Noranizan MA. Aflatoxins and ochratoxin a reduction in black and white pepper by gamma

radiation. Radiation Physics and Chemistry. 2012;**81**:1786-1788

[110] Dhanya R, Mishra BB, Khaleel KM, Cheruth AJ. Shelf life extension of fresh turmeric (*Curcumal longa* L.) using gamma radiation. Radiation Physics and Chemistry. 2009;**78**:791-795

[111] Queirol MAP, Neto JT, Arthur V, Wiendl FM, Villavicencio AH. Gamma radiation, cold and four different wrappings to preserve ginger rhizomes, Zingiber officinallis Roscoe. Radiation Physics and Chemistry. 2002;**63**:341-343

[112] Gourama H, Bullerman LB. Aspergillus flavus and aspergillus parasiticus: Aflatoxigenic fungi of concern in foods and feeds: A review. Journal of Food Protection. 1995;**58**(12):1395-1404

[113] Lancaster MC, Jenkins F, Philp JM. Toxicity associated with certain samples of groundnuts. Nature. 1961;**192**:1095-1096

[114] Hall AJ, Wild CP, Eaton DL, Groopman JD. Epidemiology of aflatoxin-related disease. In: The toxicology of aflatoxins: Human health, veterinary and agricultural. Significance. San Diego, CA, USA: Academic Press; 1994. pp. 233-258

[115] Massey TE, Stewart RK, Daniels JM, Liu L. Biochemical and molecular aspects of mammalian susceptibility to aflatoxin B-1 carcinogenicity. Proceedings of the Society for Experimental Biology and Medicine. 1995;**208**(3):213-227

[116] MFDS (Ministry of Food and Drug Safety). Aflatoxin Risk Profile. 2010. Available from: http://www.foodnara.go.kr/foodnara/board-read.do?boardNo=1309162875941&command=READ&mid=S07_05_03&boardId=series> [Accessed: March 16, 2014]

[117] Pankaj SK, Shi H, Keener KM. A review of novel physical and chemical decontamination technologies for aflatoxin in food. Trends in Food Science and Technology. 2018;**71**:73-83

[118] Peng Z, Chen L, Zhu Y, Huang Y, Hu X, Wu Q, et al. Current major degradation methods for aflatoxins: A review. Trends in Food Science & Technology. 2018;**80**:155-166

[119] Ryu D, Bianchini A, Bullerman LB. Effects of processing on mycotoxins. Stewart Postharvest Review. 2008;**6**:1-7

[120] Lee J, Her JY, Lee KG. Reduction of aflatoxins (B1, B2, G1, and G2) in soybean-based model systems. Food Chemistry. 2015;**189**:45-51

[121] Markov K, Mihaljevi'c B, Domijan AM, Pleadin J, Delaš F, Frece J. Inactivation of aflatoxigenic fungi and the reduction of aflatoxin B1 in vitro and in situ using gamma irradiation. Food Control. 2015;**54**:79-85

[122] Serra MS, Pulles MB, Mayanquer FT, Vallejo MC, Rosero MI, Ortega JM, et al. Evaluation of the use of gamma radiation for reduction of aflatoxin B1 in corn (Zea mays) used in the production of feed for broiler chickens. Journal of Agricultural Chemistry and Environment. 2018;7:21-33

[123] Mohamed NF, El-Dine RSS, Kot MAM, Saber A. Assessing the possible effect of gamma irradiation on the reduction of aflatoxin B1, and on the moisture content in some cereal grains. American Journal of Biomedical Science and Research. 2015;7:33-39

[124] Zhang ZS, Xie QF, Che LM. Effects of gamma irradiation on aflatoxin B 1 levels in soybean and on the properties of soybean and soybean oil. Applied Radiation and Isotopes. 2018;**139**:224-230

[125] Ismail A, Gonçalves BL, de Neeff DV, Ponzilacqua B, Coppa CFSC, Hintzsche H, et al. Aflatoxin in foodstuffs: Occurrence and recent advances in decontamination. Food Research International. 2018;**113**:74-85

[126] Loi M, Renaud JB, Rosini E, Pollegioni L, Vignali E, Haidukowski M, et al. Enzymatic transformation of aflatoxin B1 by Rh_DypB peroxidase and characterization of the reaction products. Chemosphere. 2020;**250**:126296

[127] Torlak E, Akata I, Erci F, Uncu AT. Use of gaseous ozone to reduce aflatoxin B1 and microorganisms in poultry feed. Journal of Stored Products Research. 2016;**68**:44-49

[128] Dogbevi MK, Vachon C, Lacroix M. Physicochemical properties of dry red kidney bean proteins and natural microflora as affected by gamma irradiation. Journal of Food Science. 1999;**64**:540-542

[129] Mallikarjunan N, Marathe S, Deshpande R, Jamdar SN, Sharma A. Influence of γ-radiation on the structure and function of soybean trypsin inhibitor. Journal of Agricultural and Food Chemistry. 2012;**60**(48):12036-12043. DOI: 10.1021/jf3038264

Chapter 5

Effect of Gamma Irradiation and/or Entomopathogenic Fungi on Some Biological Aspects of *Galleria mellonella* L. (Lepidoptera: Pyralidae)

Hussein Farid Mohamed, Samira Elsayed Mustafa El-Naggar, Mahmoud Abd-elmohsen Sweilem, Ahmed Adly Mahmoud Ibrahim and Ola Elsayed Abd Alrahman El-khawaga

Abstract

Studying the impact of gamma radiation in conjunction with the LC_{50} of the entomopathogenic fungi on a few biological characteristics of *Galleria mellonella* (L.) larvae was the goal of the current work. The effects of single doses of gamma radiation (70, 100, 125, and 150 Gy) given sequentially or combined with the LC50 of specific entomopathogenic fungi (*Beauveria bassiana, Paecilomyces lilacinus*) against the Greater wax moth. The first generation (F_1) of *G. mellonella* was examined for its larval pupal period, pupation percent, sex ratio, and adult survival percent. Gamma irradiation and LC_{50} of combined treatments of *Beauveria bassiana* and *Paecilomyces lilacinus* prolonged the larval-pupal period, while pupation, adult emergence, survival and sex ratio were lower in the combined treatment than in either treatment alone get more. The combined treatment was greater than either, in the case of fungal or irradiation treatment alone.

Keywords: entomopathogenic fungi, *galleria mellonella*, gamma irradiation, gamma rays, larvae

1. Introduction

One of the most obnoxious and significant wax pests in the world is the greater wax moth (GWM), *Galleria mellonella* L. (Lepidoptera: Pyralidae) [1–4]. In its larval stage, this larva badly harms comb wax by feeding on it. When bees neglect to tend to combs, the wax moth larvae severely damage them. Attacks are possible on combs in weak or dead colonies and storage spaces [5, 6]. The generations of virulent strains for target pests and their usage as biological control agents have attracted a lot of research attention [7, 3].

Scientists have given entomopathogenic fungi, which infect insects, a lot of thought because of their potential for biological pest control. While certain insect pathogenic fungus has narrow host ranges, others, including *Beauveria bassiana* and *Paecilomyces lilacinus*, have broad host ranges [8].

Entomopathogenic fungi (EPFs) are one of highly specialized microorganisms that could infect and grow on arthropods [9]. EPFs such as *Beauveria bassiana* and *Metarhizium anisopliae* are widely used in the control of several insects [10]. *Beauveria bassiana* (Balsamo) (Ascomycota: Hypocreales: Clavicipitaceae) is a common fungus with effects on a wide variety of insect species and is used as a biological pesticide [11]. Several studies have been published on the efficacy of EPF against *G. mellonella* [12].

The pathogenicity of the fungi against the tested larvae was increased by gamma irradiation. The two control methods may be combined to effectively eliminate the insect pest, especially in storage [1, 13].

The aim of the current work, study the effect of gamma irradiation in combination with LC_{50} of entomopathogenic fungi on some biological aspects of the 2nd instar larvae of *Galleria mellonella* (L.)

2. Materials and methods

2.1 Insect rearing and irradiation process

The greater wax moth, *G. mellonella* larvae were obtained from infested hives and reared in the Nuclear Research Center (NRC), Egyptian Atomic Energy Authority (EAEA), Egypt, Anshas area, the bio-insecticide Production Unit, Plant Protection Research Institute, Agricultural Research Center, Giza, Egypt. *G. mellonella* larvae reared on the artificial diet at a constant temperature of 30°C and 65 ± 5% relative humidity (R.H) according to ref. [14]. The irradiation process was performed using cobalt-60 gamma cell 220 located at Cyclotron project, Nuclear Research Centre, Atomic Energy Authority (Anshas). The dose rate was 0.926 kGy/hour during the experiment. Full-grown pupae of *G. mellonella* were exposed to four doses 70, 100, 125, and 150 Gy to study the effect of irradiation on some biological aspects of F_1 progeny descendant of irradiated parental male and female pupae.

Ten larvae resulting from irradiated parental males and females and transferred to clean small plastic containers and allowed to feed on an artificial diet, each treatment was replicated five times. Dead larvae were counted. The larval and pupal period, pupation, adult emergence, sex ratio, Adult survival percentage, and the resulting progeny (F_1) were determined as well.

2.2 The experimental media

Culture media were adjusted by the addition of: –10 μg/ml Dodine, 100 μg/ml Chloramphenicol, 50 μg/ml Streptomycin and 32.5 g SDA (sabouraud dextrose agar) in 500 ml distilled water "*Dodine media*" [15].

Chloramphenicol, thiabendazole and cycloheximide (CTC) medium consisting of CDA supplemented with 0.5 g/l chloramphenicol and 0.25 g/l cycloheximide) and 0.002 g/l thiabendazole [16].

2.3 Isolation of the entomopathogenic fungi

The entomopathogenic fungal, *Paecilomyces lilacinus* isolates from soils and the entomopathogenic fungal, *Beauveria bassiana* isolates from dead insects [17].

2.3.1 Identification of the fungal isolates

The fungal colonies arising on the plates were purified. The purified cultures grown on CDA and SDA (sabouraud dextrose agar) media were identified at Mycological Center (AUMC), Assiut University, Egypt.

The strain numbers for the *B. bassiana* was (AUMC 9894) and strain numbers for the *Paecilomyces lilacinus* was (AUMC 9884).

2.4 Bio efficacy of Entomopathogenic fungi against *G. mellonella* 2nd instar larvae

To ascertain whether the most virulent isolate of fungi is pathogenic to *G. mellonella* larvae, According to ref. [1], the second instar larvae were each submerged for 30 seconds in nine ml of various spore concentrations from the fungal isolates. Larvae were dipped into a 0.02% Triton X-100 solution as the control treatment [18]. After that, the treated larvae were put one by one in little plastic containers and given an artificial food to consume. All treated larvae were incubated for 12 hours during the photo phase at 30°C, 65 ± 5% relative humidity. Five times each of each treatment's batches of 10 larvae were reproduced. Daily counts of dead larvae were made. According to *Finney's* approach, the median fatal concentration and time were determined [19].

2.4.1 The investigated entomopathogenic fungi's latent effect on a few biological features of G. mellonella

Beauveria bassiana and *Paecilomyces lilacinus* were utilized to estimate biological activity in light of the high mortality rate. To investigate how Entomopathogenic fungi affect several biological characteristics, the concentrations necessary to kill 50% of larvae during the observation time (LC_{50}) were selected (**Table 1**).

To evaluate the biological activity of *G. mellonella* under laboratory conditions, four replicates of the 2nd instar larvae (20 larvae for each) were dipped into 9 ml of LC_{50} of the tested entomopathogenic fungi (*B. bassiana and P. lilacinus*) for 30 seconds. Then treated larvae were placed individually in small plastic containers and allowed to feed on the semi-synthetic diet. For the control treatment, larvae were dipped into 0.02% Triton X-100 solution Dead larvae were counted daily. Growth

Fungal isolates	Lc_{50} conidia/ml	Lc_{95} conidia/ml	slope	X^2	P value
Beauveria bassiana	1.2×10^5	1.9×10^8	0.2620 ± 0.0305	13.15	0.001
Paecilomyces lilacinus	2.3×10^5	4.3×10^{11}	0.2064 ± 0.0299	04.30	0.120

**Expressed as the LC_{50}, LC_{95}, and slope of toxicity regression lines after 10 days of dipping in different concentration.*

Table 1.
Virulence of fungal isolates against 2nd instar larvae of G. mellonella.*

parameters, namely larval – pupal period, pupation, adult emergence, sex ratio and survival (%) were recorded.

2.4.2 Effect of entomopathogenic fungi and gamma irradiation on some biological aspects of G. mellonella

To investigate the combined affects of gamma irradiation with the (LC_{50}) of various Entomopathogenic fungi (*B. bassiana* and *P. lilacinus*), three dose levels of gamma irradiation (70, 100, 125) were used. Three experimental groups were set up to explore the influence on some biological characteristics of *G. mellonella*. The first group was made up of the 70, 100, and 125Gy-irradiated F_1 larvae offspring of the irradiated Parental females as fully formed pupae. The second group was made up of the 70, 100, and 125Gy-irradiated full-grown pupae of F_1 larvae descended from irradiated parental males. Insect males and females were utilized as a control in the third group, which served as a parallel group of unirradiated insects.

The progeny of F_1 larvae of each group were fed on an artificial diet till the 2nd instar larvae, five replicates from the 2nd instar larvae (10 larvae each) were dipped into 9 ml of LC_{50} of the tested entomopathogenic fungi (*B. bassiana* and *P. lilacinus*) for 30 seconds in clean small plastic containers fitted with moist filter paper and allowed to feed on an artificial diet under laboratory condition (28 ± 2°C and 65 ± 5% relative humidity). Larval and pupal period, the percentage pupation, the percentage of adult emergence, sex ratio and the percentage of adult survival were determined.

2.5 Statistical analysis

The lethal concentration 50 was determined for established regression lines. All data obtained were analyzed using the Analysis of Variance (ANOVA) technique and the means were separated using Duncan's multiple range test ($P > 0.05$) [20].

3. Results

The latent effect of gamma irradiation on the larval and pupal periods of F_1 progeny descendant of irradiated parental males and females as full-grown pupae.

Figure 1 demonstrates how gamma radiation affected the larval and pupal periods of the greater wax moth, *Galleria mellonella*, in F_1 male offspring of the irradiated paternal males and females as full-grown pupae with four doses of 70, 100, 125, and 150 gray with an increase in dose, the average larval and pupal period lengthens noticeably.

The larval and pupal periods of the greater wax moth, *Galleria mellonella*, in F_1 male offspring of the irradiated paternal males rises to 41.73, 44.72, 45.80, and 47 days for the four doses, respectively with an increase in dose in comparison to 35.25 days for the control treatment. Also, the larval and pupal periods of the greater wax moth, *Galleria mellonella*, in F_1 male offspring of the irradiated paternal females dramatically lengthens with an increase in dose, as seen in **Figure 1**. Compared to 35.25 days for the control treatment, it rises to 42.27, 44.41, and 46.26 days for the three doses, respectively.

The combined impact of Entomopathogenic fungi and gamma irradiation on the larval and pupal periods of F_1 progeny descendant of irradiated parental males and females as full-grown pupae.

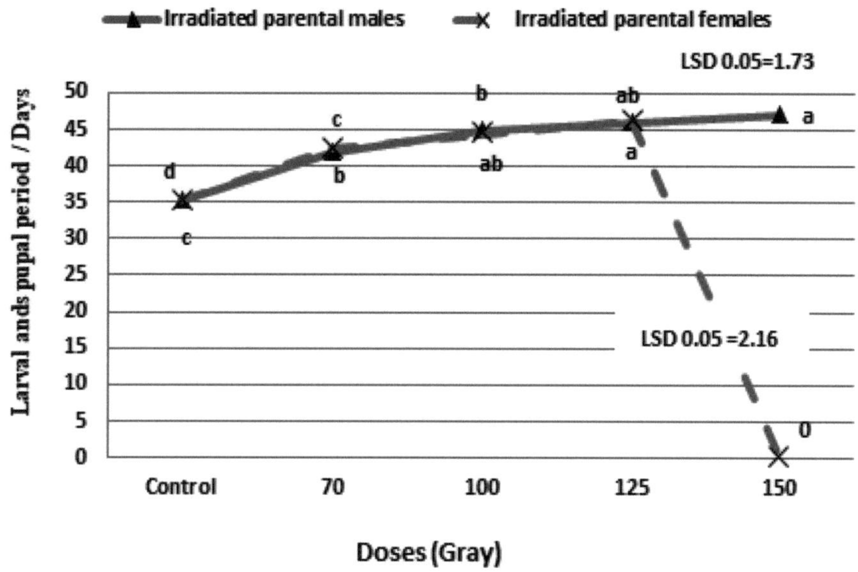

Figure 1.
*The effect of gamma irradiation on the larval and pupal periods of the greater wax moth, Galleria mellonella, in the F_1 male progeny of the irradiated parental males and females as full-grown pupae. *Means followed by the same letter in each curve (small letters) represent that are not significantly different at ($p < 0.05$).*

Figure 2 shows the impact of gamma irradiation in conjunction with the LC_{50} of the entomopathogenic fungi, *B. bassiana* and *P. lilacinus*, on the average larval and pupal period per day of *G. mellonella* descending of the irradiated parental males and females. As the irradiation dose was increased, there was a significant increase in the mean larval and pupal durations (**Figure 2**). The average larval and pupal period among F_1descendant of the irradiated parental males (**Figure 2, Hist. 1**) significantly increased from 36.22 days in the control treatment to 43.55, 46.33, and 48.10 days for *B. bassiana* combined with the doses of 70, 100, and 125 Gy, respectively.

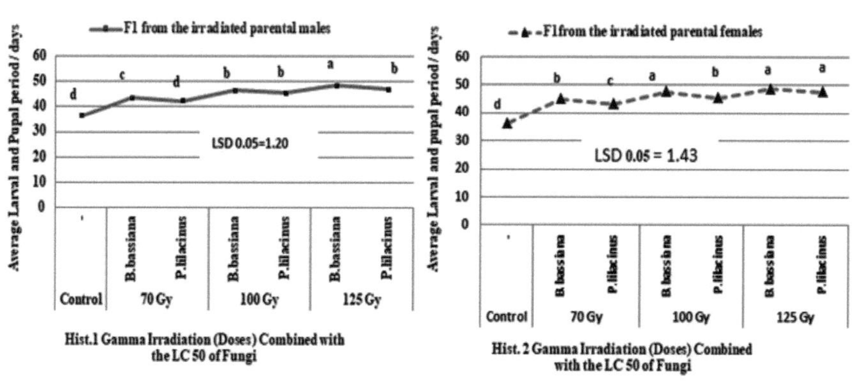

Figure 2.
*The effect of gamma irradiation combined with the LC_{50} of the entomopathogenic fungi on the average larval and pupal period per day of G. mellonella F_1 males. Hist. 1: F_1 male's descendants of irradiated parental males at three different doses: 70, 100, and 125 gray. Hist. 2: F_1 male's descendants of irradiated parental females at three different doses: 70, 100, and 125 gray. *Means followed by the same letter in each curve (small letters) represent that are not significantly different at ($p < 0.05$).*

Additionally, for *P. lilacinus* paired with the identical prior irradiation dosages, it dramatically rose to 42.14, 45.21, and 46.63 days, respectively.

The effect of gamma irradiation in combination with the (LC_{50}) of the entomopathogenic fungi, *B. bassiana* and *P. lilacinus*, is shown in **Figure 2**, **Hist. 2** with respect to the mean larval and pupal period of *G. mellonella* descending from the irradiated parental females. With increasing irradiation dose, the mean larval and pupal period increased considerably. It significantly increased from 36.22 days in the control treatment to 44.73, 47.56, and 48.70 days at 70, 100, and 125 Gy dosages combined with *B. bassiana*. Additionally, it significantly increased from 36.22 days in the control treatment to 43.27, 45.16, and 47.40 days at 70, 100, and 125 Gy, respectively, when coupled with *P. lilacinus*.

The impact of gamma radiation on the sex ratio of F_1 male's descendant of the irradiated parental males and females as full-grown pupae of the Greater wax moth, Galleria mellonella.

Table 2 shows the impact of gamma irradiation on the sex ratio of F_1 male's descendant of the irradiated parental males and females as fully grown pupae of the greater wax moth, Galleria mellonella. The average sex ratio through the irradiated parental males as full-grown pupae rose. At the four doses of 70, 100, 125, and 150 Gy, it rises to 1.20, 1.33, 1.40, and 1.60, respectively, as opposed to 1.18 for the untreated control treatment) however, this increase was not significant). As well as the sex ratio through the irradiated parental males as full-grown pupae, with the dose increase, the average sex ratio went up. At the three doses of 70, 100, and 125 Gy, it rose from 1.18 for the control treatment to 1.40, 1.60, and 1.80, respectively) also, this increase was not significant).

The combined impact of Entomopathogenic fungi and gamma irradiation on the sex ratio of F_1 male's descendant of the irradiated parental males and females as full-grown pupae of the Greater wax moth, Galleria mellonella.

Table 3 Demonstrated the relationship between the entomopathogenic fungi's (LC_{50}) impact and gamma irradiation on the *sex ratio* of *G. mellonella* descending from irradiated parental males and females. The sex ratio of *G. mellonella* descending of the irradiated parental males reveals that it was 1.4 and 1.6 in favor of males for *B. bassiana* paired with doses of 70 and 100 Gy, respectively, but was 1:1 at the dosage rate of 125 Gy as opposed to 1.13:1.0 at the control treatment. Additionally, it rose to

Doses (Gy)	Sex ratio ± SE			
	irradiated parental males as full-grown pupae		irradiated parental females as full-grown pupae	
	male	female	male	female
Control (0)	1.18 ± 0.13 a*	1	1.18 ± 0.13 a*	1
70	1.20 ± 0.08 a	1	1.40 ± 0.08 a	1
100	1.33 ± 0.09 a	1	1.60 ± 0.09 a	1
125	1.40 ± 0.18 a	1	1.80 ± 0.18 a	1
150	1.60 ± 0.24 a	1	1.18 ± 0.13 a	1
LSD 0.05	0.47		0.47	

**Means followed by the same letter in each column (small letters) represent that are not significantly different at ($p < 0.05$).*

Table 2.
Effect of gamma irradiation on the sex ratio, of the greater wax moth, galleria mellonella, among F_1 male's progeny of the irradiated parental males and females as full-grown pupae.

Radiation Doses (Gy)	Fungi	Sex ratio ± SE			
		irradiated parental males as full-grown pupae		irradiated parental females as full-grown pupae	
		Male ± SE	**Female**	**Male ± SE**	**Female**
Control	—	1.13 ± 0.11 a*	1	1.1 ± 0.11 a	1
70 Gy	*B. bassiana*	1.40 ± 0.48 a	1	1.5 ± 0.44 a	1
	P. lilacinus	1.33 ± 0.27 a	1	1.7 ± 0.43 a	1
100 Gy	*B. bassiana*	1.60 ± 0.24 a	1	1.0 ± 0.31 a	1
	P. lilacinus	1.30 ± 0.30 a	1	1.2 ± 0.37 a	1
125 Gy	*B. bassiana*	1.00 ± 0.31 a	1	0.6 ± 0.40 a	1
	P. lilacinus	1.20 ± 0.20 a	1	0.8 ± 0.37 a	1
LSD		0.904		1.06	

**Means followed by the same letter in each column represent that are not significantly different at ($p < 0.05$).*

Table 3.
The effect of gamma irradiation and the entomopathogenic fungi's (LC_{50}) on the sex ratio of G. mellonella descending of the irradiated parental males and females.

1.33, 1.30, and 1.20 following *P. lilacinus* treatment in conjunction with the identical dosages of prior irradiation, respectively.

The sex ratio of *G. mellonella* descending of the irradiated parental females also (**Table 3**) reveals that at the two doses of 70 Gy and 100 Gy mixed with *B. bassiana*, was in favor of males (1.5 and 1.0, respectively). However, it was decreased from 1.1 in the control treatment to 0.6 with a dose of 125 Gy. At the two doses of 70 Gy and 100 Gy mixed with *P. lilaceous*, it also climbed to 1.7 and 1.2, respectively. At 125 Gy, however, it decreased to 0.8. (All data in **Table 3** was not significant).

The impact of gamma radiation on the percentages of pupation, emergence, and survival of F_1 male's descendant of the irradiated parental males and females as full-grown pupae of the Greater wax moth, Galleria mellonella.

Figure 3 shows the percentages of pupation, emergence, and survival of *G. mellonella* in F_1 male descendants of irradiated parental males at four different doses: 70, 100, 125, and 150 gray (**Figure 3, Hist. 1**). The percentage of pupation and the rise in dosages were negatively correlated. When compared to 92% for the control treatment, the percentage of pupation significantly drops to 78, 74, 68, and 60% at the four doses of 70, 100, 125, and 150Gy, respectively. Meanwhile, the percentage of the F_1 generation's adult emergence greatly drops for all treatments. It drops from 93.55% for the control treatment to 82.69, 80.71, 70.23, and 56.76% at the four doses of 70, 100, 125, and 150Gy, respectively (**Figure 3, Hist. 1**). Additionally, **Figure 3, Hist. 1** shows that as the dose was increased, the proportion of larvae that reached the adult stage considerably dropped (% Survival). At the four doses of 70, 100, 125, and 150Gy, the survival percentage fell to 66, 60, 48, and 34, respectively, from 86% for the control treatment.

The percentages of pupation, emergence, and survival of *G. mellonella* in F_1 male descendants of irradiated parental females at four different doses: 70, 100, 125, and 150 gray, It has been stated in the **Figure 3, Hist. 2**. At the four dosages of 70, 100, and 125 Gy, respectively, the percentage of pupation significantly decreased to 62, 54, and 38% compared to 92% for the control treatment, while, the parental females that received the dose of 150 Gy did not produce any offspring.

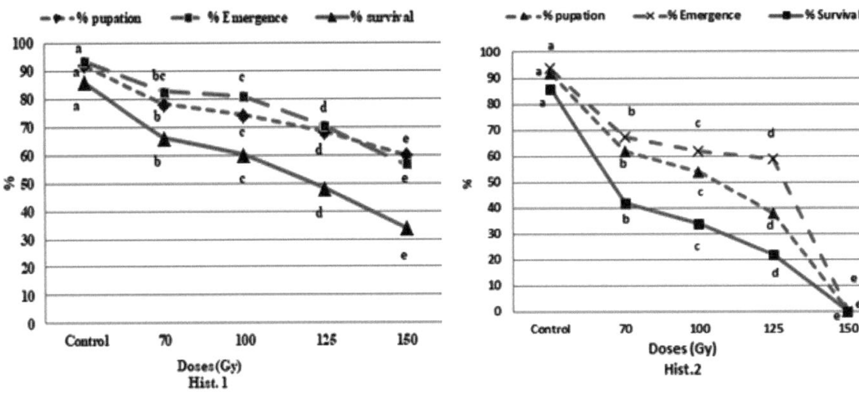

Figure 3.
*A negative association between the increase in dosages and the percentage of pupation, emergence, and survival of G. mellonella. Hist. 1: F_1 males of G. mellonella descendants of irradiated parental males at four different doses: 70, 100, 125, and 150 gray. Hist. 2: F_1 males of G. mellonella descendants of irradiated parental females at three different doses: 70, 100, and 125 gray. *Means followed by the same letter in each curve (small letters) represent that are not significantly different at ($p < 0.05$).*

At all treatments, there was a considerable reduction in the F_1 generation's (F_1 females') adult emergence percentage. It drops from 93.55% for the control treatment to 67.23, 62.00, and 58.66% for the three dosages of 70, 100, and 125 Gy, respectively (**Figure 3, Hist. 2**) while, the parental females that received the dose of 150 Gy did not produce any offspring as mentioned previously.

Additionally, a significant decrease in the percentage of larvae that reached the adult stage (**Figure 3, Hist. 2**) was seen as the dose was increased. In comparison to the control treatment, the percentage of survivorship reduced to 42, 34, and 22% at the three doses of 70, 100, and 125 Gy, respectively.

The combined impact of Entomopathogenic fungi and gamma radiation on the percentages of pupation, emergence, and survival of the F_1 males of the Greater wax moth, Galleria mellonella descendant of irradiated parental males and females.

Figure 4 illustrates how gamma radiation and the entomopathogenic fungi's LC_{50} affect the proportion of pupation, emergence, and survival of the F_1 males of the Greater wax moth, Galleria mellonella descendant of irradiated parental males and females.

The effects of Entomopathogenic fungus and gamma radiation on F_1 males of *G. mellonella* descendants of the irradiated parental males with doses of 70, 100, and 125 Gy are explained in **Figure 4, Hist. 1**. With increasing irradiation dose, the percentage of pupation was dramatically reduced, and treatments on *B. bassiana* showed greater declines than those on *P. lilacinus.* When combined with doses of 70, 100, and 125 Gy, respectively, it significantly decreased for *B. bassiana* to 42, 30, and 22%, respectively, compared to 80% in the control treatment, while it significantly decreased for *P. lilacinus* to 50, 38, and 28%, respectively, when combined with the same previous irradiation doses.

Also, with increasing irradiation dose, the percentage of pupation among F_1 males of *G. mellonella* descendants of irradiated parental males (**Figure 4, Hist. 1**) was significantly reduced, and treatments on *B. bassiana* showed greater declines than those on *P. lilacinus* when combined with doses of 70, 100, and 125 Gy, respectively, it significantly decreased for *B. bassiana* to 42, 30, and 22%, respectively, compared to 80% in the control treatment, while it significantly decreased for *P. lilacinus* to 50, 38, and 28%, respectively, when combined with the same previous irradiation doses.

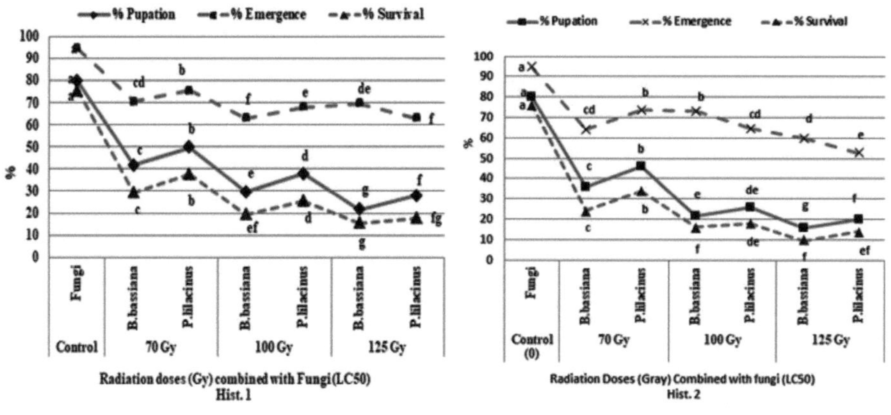

Figure 4.
*The effected of gamma irradiation and the entomopathogenic fungi's (LC_{50}) on the percentage of pupation, emergence, and the adult survival of G. mellonella. Hist. 1: F_1 males of G. mellonella descendants of irradiated parental males. Hist. 2: F_1 males of G. mellonella descendants of irradiated parental females. *Means followed by the same letter in each curve (small letters) represent that are not significantly different at ($p < 0.05$).*

In comparison to the control treatment, the percentage of adult emergence for *B. bassiana* combined with doses of 70, 100, and 125 Gy was considerably reduced to 71, 63.33, and 70%, respectively. Additionally, for *P. lilacinus* paired with the identical prior irradiation dosages, it significantly decreased to 75, 70, and 63.33%, respectively (**Figure 4, Hist. 1**).

Figure 4, Hist. 1 also shows that as the irradiation dose was increased, the proportion of survival fell. In comparison to the control treatment, it was significantly reduced, dropping to 30, 20, and 16% for *B. bassiana* paired with dosages of 70, 100, and 125 Gy, respectively. Additionally, for *P. lilacinus* treatment paired with the same prior irradiation dosages, it significantly decreased to 38, 26 and 18%, respectively.

The percentage of pupation, emergence, and adult survival of *G. mellonella* descended from irradiated parental females is shown in **Figure 4, Hist. 2** to be affected by gamma irradiation and the entomopathogenic fungi's (LC_{50}). The percentage of pupation was significantly decreased with the increasing irradiation dose. It significantly decreases to 36, 22 and 16% with the three doses 70, 100 and 125 Gy combined with *B. bassiana* respectively, compared to 80% in the control treatment. Also, it significantly decreased to 46, 26 and 20% with the three doses 70, 100 and 125 Gy combined with *P. lilacinus* respectively.

Additionally, **Figure 4, Hist. 2** shows that for all treatments in *B. bassiana* and *P. lilacinus*, the percentage of emergence decreases when the irradiation dose is increased. In comparison to the control treatment, it considerably dropped from 95% to 64.33, 73.33, and 60.00% at the three doses of 70, 100, and 125 Gy coupled with *B. bassiana*, respectively. While it significantly decreases from 95% in the control treatment to 74.00, 65.00, and 53.000% at the three dosages of 70, 100, and 125 Gy coupled with *P. lilacinus*, respectively.

Also, **Figure 4, Hist. 2** shows that as the irradiation dose are increased, the percentage of adult survival decreases and it decrease more in *B. bassiana* than in *P. lilacinus*. When joined with *B. bassiana* at the three doses of 70, 100, and 125 Gy, it was significantly reduced from 76% in the control treatment to 24, 16, and 10%, respectively. While at the three doses of 70, 100, and 125 Gy joined with *P. lilacinus*, it significantly decreases to 34, 18 and 14%, respectively.

4. Discussions

Studies have been done on the biological activity (larval and pupal period, pupation, adult emergence, sex ratio, and survival) of *G. mellonella* second instar larvae fed with LC_{50} amounts of *B. bassiana* and *P. lilacinus*. According to the data, the entomopathogenic fungi's LC_{50} significantly lengthened the larval and pupal stages compared to the control. These findings support those of ref. [21]. The screening of isolated fungi's crude extracts for *G. mellonella* was acknowledged in 2012. Different fungal crude extracts of isolates were used, with varying effects on the percentages of larval and adult mortality. The Siwa and El-Farafra isolates of *Beauveria bassiana* had an impact on *G. mellonella* larvae, causing total mortality percentages of 82.48 and 74.22%, respectively, in comparison to the control treatment of 0%. The results showed that *B. bassiana* (El-Farafra and Siwa Oasis isolates) had a substantial impact on *G. mellonella* (larval and pupal weights). With increasing *B. bassiana* crude extract content, the percentage of larval mortality rose (El-Farafra and Siwa isolates). Similarly, ref. [22] discovered that *B. bassiana* dramatically shortened the *S. litura* larval period, and that this shortened period was observed in treated larvae of the second and fourth instars. Additionally, the LC_{50} of *B. bassiana* and *P. lilacinus* decreased *G. mellonella* pupation, adult emergence, and the percent survival. Similar to that, the current experiment supports [23] findings that *B. bassiana* at 10^8 conidia /ml reduces *G. mellonella* pupation and adult emergence to 20 and 13.33%, respectively. Additionally, ref. [24] discovered that *B. bassiana* caused the least amount of pupation and adult emergence of *S. litura* at spore concentrations of 2.4×10^7 and 2.4×10^4 conidia/ml, respectively. In contrast, ref. [23], who noted that when *G. mellonella* larvae in their fourth instar were treated with *B. bassiana*, the male to female ratio was higher. The sex ratio of the F_1 progeny was also biased in favor of the males. In their studies on *Spodoptera litura* in 2002, ref. [25] found that the sex ratio was biased in favor of males. Our findings demonstrated that the combined effect decreased the percentage of pupation, adult emergence, and survival; it also lengthened the larval and pupal duration. The F_1 generation's gender ratio likewise shifted more in favor of males. The results of the current investigation showed that gamma irradiation and entomopathogenic fungi worked together to produce effects that were more notable than any of them working alone. The effectiveness of *B. bassiana* increased, especially when the gamma irradiation dose was increased, where no adults were produced with both the fungal concentrations and 150 Gy gamma irradiation dose, according to ref. [23] study using *B. bassiana* at 10^4 and 10^8 spores per milliliter combined with different doses of gamma irradiation (50, 100, and 150). According to ref. [1] gamma irradiation increased the fungi's pathogenicity toward the studied larvae. The two pest-control instruments working together could effectively eliminate insects, especially in storage. Ref. [3] Investigated the effectiveness of *Beauveria bassiana* (Biovar), *Trichoderma album* (Biozed), and *Metarhizium anisopliae* (Bioranza) on *Galleria mellonella* (L.) greater wax moth larvae (Lepidoptera: galleriidae). They stated that some biological characteristics of *Galleria mellonella* larvae are significantly impacted by the three tested biocides.

5. Conclusions

In general, the results indicated that females were more radiosensitive to gamma irradiation than males. The F_1 progeny resulting from the irradiated full-grown male pupae was more sterile than the irradiated P_1 males and the F_1 males were usually more sterile than the F_1 females. The entomopathogenic fungi, *B. bassiana* and *P. lilacinus* were the most virulent isolates causing the highest adult longevity against 2nd instar larvae of *Galleria mellonella* and had the lowest LC_{50} values. Full-grown male and female pupae irradiated with different doses of gamma irradiation (70, 100, 125 and 150 Gy) showed delayed effects on the different biological aspects of *G. mellonella* females were more radiosensitive than males. The sterilizing dose for males was above 150 Gy, while the sterilizing dose for females was 150 Gy.

The second instar larvae of *Galleria mellonella* descending of the irradiated parental male or female pupae with 70, 100 and 125 Gy then treated with the LC_{50} of *B. bassiana* and *P. lilacinus* adversely affected on the different biological aspects. The combined treatment was greater than either, in the case of fungal or irradiation treatment alone.

Authors' contributions

The Manuscript writing was done by all authors. Tables and reference settings were done by all authors. Then, final manuscript was read and approved by all authors. All authors read and approved the final manuscript.

Funding

No funding.

Conflict of interest

The authors declare no conflict of interest.

Consent for publication

The authors consent for publication.

Availability of data and materials

Data supporting the conclusions of this article are presented in the main Manuscript.

Ethics approval and consent to participate

The authors declare that they have ethics approval and consent to participate.

Author details

Hussein Farid Mohamed[1]*, Samira Elsayed Mustafa El-Naggar[1], Mahmoud Abd-elmohsen Sweilem[2], Ahmed Adly Mahmoud Ibrahim[3] and Ola Elsayed Abd Alrahman El-khawaga[1]

1 Biological Application Department, Nuclear Research Centre, Atomic Energy Authority, Egypt

2 Botany Department, Faculty of Science, Benha University, Egypt

3 Plant Protection Research Institute, Agriculture Research Center, Cairo, Egypt

*Address all correspondence to: husseinmohamed29@yahoo.com

References

[1] Hussein FM, Thanaa MS, Samira EME, Mahmoud AES, Ahmed AMI, Ola EAAEE. Effect of gamma irradiation and/or certain entomopathogenic fungi on the larval mortality of *galleria mellonella* L. Egyptian Journal of Biological Pest Control. 2018;**28**(95):2-8. DOI: 10.1186/s41938-018-0099-z

[2] Shoukry IF, Ahmed FA, Khater KS, El-lakwah SF, Abd-Elmonem HM. Evaluation of the effectiveness of some Entomopathogenic fungi on the greater wax moth larvae, *galleria mellonella* (L.) (Lepidoptera: Galleriidae). Egyptian Academic Journal of Biological Sciences. 2019;**12**(4):41-55. ISSN 1687-8809. Available from: http://eajbsa.journals.ekb.eg/

[3] Słowińska M, Nynca J, Bąk B, Wilde J, Siuda M, Ciereszko A. 2D-DIGE proteomic analysis reveals changes in haemolymph proteome of 1-day-old honey bee (*Apis mellifera*) workers in response to infection with Varroa destructor mites. Apidologie. 2019;**50**:632-656. DOI: 10.1007/s13592-019-00674-z

[4] Kumar G, Khan MS. Ecofriendly Management of Greater wax Moth (*galleria mellonella*) infesting combs under storage. International Journal of Pure & Applied Bioscience. 2020;**8**(4):237-245. DOI: 10.18782/2582-2845.7632

[5] Saikia SS, Borah BK, Baruah G, Rokozeno DMK. Characterization of the gut microbes of greater wax moth (*galleria mellonella* Linnaeus) shows presence of potential polymer degraders. Folia Microbiologica. 2022;**67**:133-140. DOI: 10.1007/s12223-021-00925-6

[6] Firacative C, Khan A, Duan S, Ferreira-Paim K, Leemon D, Meyer W. Rearing and maintenance of *galleria mellonella* and its application to study fungal virulence. Journal of Fungi. 2020;**6**(3):130-141. DOI: 10.3390/jof6030130

[7] Nazir T, Basit A, Hanan A, Majeed MZ, Dewen Q. In Vitro Pathogenicity of Some Entomopathogenic Fungal Strains against Green Peach Aphid *Myzus persicae* (Homoptera:Aphididae). Agronomy. 2019;**9**(7):1-12. DOI: 10.3390/agronomy9010007

[8] El Husseini MM. Pathogenicity of nuclear polyhedrosis virus to *galleria mellonella L* (*Lepidoptera: Pyralidae*) and its control on stored beeswax foundations. Egyptian Journal Of Biological Pest Control. 2020;**30**:101-106. DOI: 10.1186/s41938-020-00302-4

[9] Vega FE. The use of fungal entomopathogens as endophytes in biological control: A review. Mycologia. 2018;**110**(1):4-30. DOI: 10.1080/00275514.2017.1418578

[10] Magan N. Fungi as Biocontrol Agents: Progress, Problems and Potential. Plant Pathology. Vol. 51. 2002. pp. 518-521

[11] Xiong Q, Xiea Y, Zhua Y, Xuea J, Li J, Fan R. Morphological and ultrastructural characterization of *Carposina sasakii* larvae (*Lepidoptera: Carposinidae*) infected by *Beauveria bassiana* (*Ascomycota: Hypocreales: Clavicipitaceae*). Micron. 2013;**44**:303-311. DOI: 10.1016/j.micron.2012.08.002

[12] Abou-Shaara HF. Effects of the fungus, *Beauveria bassiana*, on the larval development of the greater wax moth, *Galleria mellonella*, (Lepidoptera: Pyralidae) under laboratory conditions. Journal of Apicultural Research. 2020;**35**(1):81-84. DOI: 10.17519/apiculture.2020.04.35.1.81

[13] Gencer D, Bayramoğlu Z. Characterization and pathogenicity of *Beauveria bassiana* strains isolated from *galleria mellonella* L. (Lepidoptera: Pyralidae) in Turkey. Egyptian Journal of Biological Pest Control. 2022;**32**(99):1-7. DOI: 10.1186/s41938-022-00599-3

[14] Metwally HM, Hafez GA, Hussein MA, Hussein MA, Salem HA, Saleh MME. Low cost artificial diet for rearing the greater wax moth, *galleria mellonella* L.(Lepidoptera: Pyralidae) as a host for entomopathogenic nematodes. Egyptian Journal of Biological Pest Control. Egyptian Society for Biological Control of Pests. 2012;**22**(1):15-17. Available from: http://www. esbcp.org/index.asp. https://www. researchgate.net/publication/286657015

[15] Shin TY, Choi JB, Bae SM, Koo HN, Woo SD. Study on selective media for isolation of entomopathogenic fungi. International Journal of Industrial Entomology. 2010;**20**(1):7-12. Available from: https://www.researchgate.net/publication/264033136

[16] Fernandes EKK, Keyser CA, Rangel DEN, Foster RN, Roberts DW. CTC medium: A novel dodine-free selective medium for isolating entomopathogenic fungi, especially *Metarhizium acridum*, from soil. Biological Control. 2010;**54**:197-205. DOI: 10.1016/j.biocontrol.2010.05.009

[17] El-khawaga OEAA. Effect of Gamma Irradiation and some Fungi on the Greater Wax Moth, Galleria Mellonella (L). [M Sc Thesis]. Faculty of Science, Benha University; 2017

[18] Ibrahim AA, Saneya RMF, Mohamed SA. Combined effect of *Beauveria Bassiana* (Bals.) and gamma irradiation on potato tuber moth *Phthorimaea Operculella* (Zeller). Egypt Journal of Agricultural Research. 2019;**97**(2):595-570. DOI: 10.21608/ejar.2019.152538

[19] Finney DJ. Probit Analysis. 3rd ed. London, UK: Cambridge University Press; 1971. p. 318. DOI: 10.1002/bimj.19720140111

[20] Steel RGD, Torrie JH. Principle and Procedures of 2nd ed. New York: McGraw-Hill book Co.; 1980

[21] Abd El-Ghany TM, El-Sheikh HH, El-Rahman GA, El-Nasser AM. Biodiversity of entomopathogenic fungi in new cultivated soil with their using to control of *galleria mellonella*. International Journal of Current Research and Review. 2012;**4**(24):17-31

[22] Kaur S, Kaur HP, Kaur K, Kaur A. Effect of different concentrations of *Beauveria bassiana* on development and reproductive potential of *Spodoptera litura* (Fabricius). Journal of Biopesticides. 2011;**4**(2):161-168

[23] El-Sinary NH, Rizk SA. Entomopathogenic fungus, *Beauveria bassiana (*Bals) and gamma irradiation efficiency against the Greater Wax Moth, *Galleria mellonella* (L). American-Eurasian Journal of Scientific Research. IDOSI Publications. 2007;**2**(1):13-18; ISSN 1818-6785

[24] Malarvannan S, Murali PD, Shanthakumar SP, Prabavathy VR, Nair S. Laboratory evaluation of the entomopathogenic fungi, *Beauveria bassiana* against the tobacco caterpillar, *Spodoptera litura* Fabricius (Noctuidae: Lepidoptera). Journal of Biopesticides. 2010;**3**(1):126-131. © JBiopest. 112. Corpus ID: 55483953

[25] Ramesh K, Garg AK, Seth RK. Interaction of substerilizing gamma radiation and thiodicarb treatment for management of the tobacco caterpillar. *Spodoptera litura*. Phytoparasitica. 2002;**30**(1):7-17